Managing the
Behaviour of Animals

Managing the Behaviour of Animals

Edited by

PAT MONAGHAN

Senior Lecturer,
Department of Zoology, University of Glasgow

and

DAVID WOOD-GUSH

Honorary Professor,
Department of Agriculture, University of Edinburgh

BOWLING GREEN STATE UNIVERSITY LIBRARIES

CHAPMAN AND HALL

LONDON · NEW YORK · TOKYO · MELBOURNE · MADRAS

UK	Chapman and Hall, 2–6 Boundary Row, London SE1 8HN
USA	Chapman and Hall, 29 West 35th Street, New York NY10001
JAPAN	Chapman and Hall Japan, Thomson Publishing Japan, Hirakawacho Nemoto Building, 7F, 1–7–11 Hirakawa-cho, Chiyoda-ku, Tokyo 102
AUSTRALIA	Chapman and Hall Australia, Thomas Nelson Australia, 480 La Trobe Street, PO Box 4725, Melbourne 3000
INDIA	Chapman and Hall India, R. Seshadri, 32 Second Main Road, CIT East, Madras 600 035

First edition 1990

© 1990 P. Monaghan and D. Wood-Gush

Typeset in 10/12pt Cheltenham Book by Leaper & Gard Ltd, Bristol
Printed in Great Britain at the University Press, Cambridge

ISBN 0 412 299801

British Library Cataloguing in Publication Data

Managing the behaviour of animals.
    1. Livestock. Management
    I. Monaghan, P. (Patricia)   II. Wood-Gush, D.G.M. (David Grainger Marcus)
    636.083

    ISBN 0–412–29980–1

Library of Congress Cataloging in Publication Data

Managing the behaviour of animals/edited by P. Monaghan
    and D.G.M. Wood-Gush. — 1st ed.
        p.   cm.
    Includes bibliographical references.
    ISBN 0-412-29980-1
    1. Animal behaviour. 2. Animal culture. 3. Animal welfare.
    I. Monaghan, P. (Pat), 1951–   . II. Wood-Gush, D. G. M. (David Grainger Marcus)
    QL751.M2198   1990
    591.51—dc20   90-1819
                                                   CIP

Contents

Contributors

IAN J.H. DUNCAN
Department of Animal and Poultry Science, University of Guelph, Ontario

VALERIE O'FARRELL
Department of Veterinary Clinical Studies, Royal School of Veterinary Studies, University of Edinburgh

PETER W. GREIG-SMITH
ADAS Central Science Laboratory, Ministry of Agriculture, Fisheries and Food, Surrey

IAN R. INGLIS
Ministry of Agriculture, Fisheries and Food, Guildford, Surrey

NEIL B. METCALFE
Department of Zoology, University of Glasgow

TREVOR B. POOLE
Universities Federation for Animal Welfare, Potters Bar, Herts

DAVID B. SHEPHERD
Ministry of Agriculture, Fisheries and Food, Guildford, Surrey

1

Introduction

Our lives are closely bound up with those of other animals, with which we interact at many different levels. We exploit other animals for food, transport, raw materials and even companionship; we compete with them for space and resources, marvel at their beauty and defend ourselves against diseases they may cause us. Consequently, there are many situations in which we need to manage the behaviour of animals.

Ethology is a branch of the science of Biology, and is concerned with explaining the causes and functions of behaviour. As such, it has an important contribution to make to the solving of problems that we face in our interactions with other animals. In addition to providing factual information on how animals behave, ethology gives us accurate methods for describing and analysing behaviour and a theoretical framework to enable us to understand and to predict the response of an animal to a particular situation (Huntingford, 1984; Monaghan, 1984). With the advent of socio-biology and behavioural ecology (Wilson, 1975; Krebs and Davies, 1978), the main thrust of ethological research has been towards studying the function of behaviour based on the assumption that behaviour has been shaped by natural selection and thereby has adaptive significance. On the other hand, as M. Dawkins stressed at the International Ethological Congress in 1987, ethologists working on the management of animals (usually termed 'applied ethologists') have been primarily concerned with the causation of behaviour, the underlying mechanisms which determine *how* the animal behaves. This is particularly obvious in work with domesticated animals and animals in zoos where confinement can create severe welfare problems. However, to manage animal behaviour successfully, we need to take account of both the adaptive significance of behaviour and the proximate mechanisms which govern its expression.

Studies of the causes of behaviour are very much concerned with explaining why the behaviour of an animal may change from one moment to the next, and thereby the understanding of how the motivational state of an animal is influenced by changes in its internal and external world. Early models of motivation, largely based on the concept of the building up of some driving force within the animal, have been replaced by models which take more account of the importance of environmental stimuli (see Huntingford, 1984 and Dawkins, 1986 for detailed discussions). As pointed out by Wood-Gush (1973), motivational theory is extremely important for understanding and dealing with the problems of the welfare of captive animals. In addressing the problem of frustration in agricultural stock he compared the motivational model proposed by Lorenz, which is based on the accumulation of so-called 'action-specific energy', with the model proposed by Hinde (1969) in relation to human aggression. In Hinde's model a particular motivational state will be partly dependent on internal factors which may vary periodically in strength, but which do not accumulate over time. The overt expression of behaviour is dependent on external stimuli. Thus, in the case of aggression, if the relevant environmental stimuli are removed, the behaviour will not occur. In contrast, with the Lorenz-type model the probability of the occurrence of different types of behaviour increases with time since they were last performed. The difference between these types of models can be illustrated by considering whether a bird which has not flown for some time develops an increasing motivation to fly, irrespective of the presence of external stimuli which might elicit a flying response. If this were the case, then housing systems of captive species would have to be designed to enable the complete behavioural repertoire to be performed without any risk to the animals themselves. While such an internally-driven view of the causation of behaviour is no longer accepted, it has been pointed out that some behavioural patterns, such as sleeping, grooming and dust bathing, do seem to fit in with the idea of an accumulating need to perform them (Wood-Gush, 1973, Vestergaard, 1982). It is likely that the extent to which internal and external factors govern the expression of a behaviour varies with the nature of the behaviour involved. This has been developed by Hughes and Duncan (1988) who cite a number of examples of captive animals performing unnecessary appetitive behaviour, and suggest that in certain cases the actual *performance* of a behaviour is important to the animal's welfare. However, they are wary of using the word 'need' because this suggests an all-or-none state. These problems make it clear how important an understanding of the causation of behaviour is to the welfare of captive animals.

In addition to the causes and functions of behaviour, ethology is also concerned with its evolution and development. Domestication is a form of micro-evolution and the behavioural characteristics that may enhance the process have been listed by Hale (1969) and are shown in Table 1.1. Such studies are valuable for the evaluation of other

Table 1 Behavioural characteristics pre-adapting species to domestication (From Hale 1969)

Favourable Characteristics	*Unfavourable Characteristics*
1. Group structure	
a. Large social group (flock, herd, pack), true leadership	a. Family groupings
b. Hierarchical group structure	b. Territorial structure
c. Males affiliated with female group	c. Males form separate groups
2. Sexual behaviour	
a. Promiscuous matings	a. Pair-bond matings
b. Clear dominance relationships	b. Male must establish dominance or appease female
c. Sexual signals provided by movements or posture	c. Sexual signals provided by colour markings or morphological structures
3. Parent–young interactions	
a. Critical period in development of species-bond (imprinting, etc.)	a. Species-bond established on basis of species characteristics
b. Female accepts other young soon after parturition or hatching	b. Young accepted on basis of species characteristics (colour patterns)
c. Precocial development of young	c. Altricial development of young
4. Responses to man	
a. Short flight distance to man	a. Extreme wariness and long flight distance
b. Least disturbed by human ubiquity and extraneous activities	b. Easily disturbed by man or sudden changes in environment
5. Other behavioural characteristics	
a. Catholic dietary habits (including scavengers)	a. Specialized dietary habits
b. Adapt to a wide range of environmental conditions	b. Fixed or unique habitat needs
c. Limited agility	c. Extreme agility

species for possible domestication and for the retention of 'wild' traits when stocks of captive bred animals are to be re-introduced into the wild to replenish endangered populations (Price, 1984). In almost all spheres of applied ethology a knowledge of how behaviour develops is essential. The effects of rearing conditions on social and foraging skills are very important to the management and welfare of animals in captivity, and again also in breeding programmes where the animals are to be returned to the wild. The learning capacities of animals and the existence of sensitive periods for sexual or filial imprinting need to be taken into account. Learning capacities are also important in pest control and the design of feeding regimes.

The following chapters are organized into three main sections. Part One deals with broad categories of behaviour which we frequently need to manipulate, that is foraging, communication and social behaviour. Many of the problems we encounter in managing the behaviour of animals involve controlling what they feed on, where they forage and how much they eat. Chapter 2 illustrates how current foraging theories can be used to manipulate these foraging decisions. That working on applied problems can highlight the limitations of current theories is demonstrated clearly here; foraging models concerned only with the economics of energy intake are too simplistic, and the palatability of the food and learning capacities of the animals need to be taken into account. In managing behaviour, whether to control pests, improve productivity or welfare or to conserve endangered species, the social behaviour and organization of the animals is an important variable which must be taken into account. This is considered in Chapter 3, with respect to social spacing, competition, group structure, early experience and mating patterns. Animals convey information to each other by means of complex signals, which may be tactile, acoustic, olfactory, visual or even electric in form. Chapter 4 examines the extent to which we can exploit such intra- or inter-specific communication to manipulate behaviour, and possibly enhance the effectiveness of the signal by altering its design. This chapter emphasizes the need to understand learning theory to use signals in this way, both in terms of the type of signal to use and how to apply it.

The remaining two parts deal with specific kinds of problems. Part Two is concerned with managing behaviour so as to maximize productivity in commercially reared species. Aquaculture is a growth industry in many parts of the world, but all too often very little account is taken of the behaviour of the animals involved. Chapter 5, which concentrates on fish, demonstrates that maximizing productivity is not simply a matter of packing fish into a tank and throwing in food.

Much more attention needs to be given to the behaviour of the fish, and particularly to factors affecting diet choice, feeding motivation, foraging and competitive abilities and life history strategies. The same holds true for the more traditional farmed animals such as cows, pigs, sheep and hens. Behavioural problems in such agricultural species are dealt with in Chapter 6. For these animals we have the additional problems associated with mating and the rearing of young under intensive farming regimes. Problems relating to disease recognition and the control of agricultural pests are also considered.

The last two chapters relate to problems associated with the welfare of animals we keep in captivity. Chapter 7 tackles the general issues of what we mean by 'welfare' and 'suffering', and how we can measure these. The difficult ethical problems involved in making decisions which relate to animal welfare issues are also discussed. This chapter illustrates the usefulness of behavioural measures as an indicator of welfare and underlines the need to understand the mechanisms underlying behaviour if problems such as the monotonous repetition of particular behaviours by captive animals (known as stereotypies) are to be solved. Finally, Chapter 8 deals with the welfare of animals that we keep mainly for companionship, particularly cats and dogs. The development of inappropriate relationships between owner and pet are at the root of many behavioural problems with these animals and this chapter outlines the ways in which an understanding of the causes and functions of particular behaviours can be used to control their occurrence.

Of course, humans have exploited, manipulated and managed the behaviour of animals over many centuries. Witness the control of the shepherd over his sheep dog, the domestication of cattle, the astonishing communication between bird and man in the complex symbiosis between the greater honeyguide (*Indicator indicator*) and the nomadic Boran people of Kenya, recently investigated in detail (Isak and Reyer, 1989). But witness too the catalogue of failures, the pitiful sight of a panther pacing its cage in a zoo, the debeaked chicken crowded in a battery cage, the blowing up of hundreds of thousands of birds in fruitless attempts to protect crops. The management of animal behaviour in the past has relied too much on unsystematic observations and superstition, with no real understanding of why problems were sometimes solved and why they were not, or indeed of why they arose in the first place. This book does not cover all of the problems we have in managing behaviour, nor does it offer universal solutions. However, the main point that the various chapters illustrate is that an understanding of the methods, theories and facts provided by the scientific discipline of ethology makes an essential contribution to the successful and acceptable management of animal behaviour.

REFERENCES

Dawkins, M.S. (1986) *Unravelling Animal Behaviour*, Longman, Harlow.

Hale, E.B. (1969) Domestication and the evolution of behaviour, in *The Behaviour of Domestic Animals*, 2nd edn (ed. E.S.E. Hafez), Ballière Tindall, London, pp. 22–42.

Hinde, R.A. (1969) The bases of aggression in animals. *J. Psychosomat. Res.*, **13**, 213–9.

Hughes, B.O. and Duncan, I.J.H. (1988) The notion of ethological need models of motivation and animal welfare. *Anim. Behav.*, **36**, 1696–707.

Huntingford, F.A. (1984) *The Study of Animal Behaviour*, Chapman and Hall, London.

Isak, H.A. and Reyer, H.U. (1989) Honey guides and honey gatherers: interspecific communication in a symbiotic relationship. *Science*, **243**, 1343–6.

Krebs, J.R. and Davies, N.B. (eds) (1978) *Behavioural Ecology: An Evolutionary Approach*, Blackwell Scientific Publications, Oxford.

Monaghan, P. (1984) Applied ethology. *Anim. Behav.*, **32**, 908–15.

Price, E.O. (1984) Behavioural aspects of animal domestication. *Quart. Rev. Biol.*, **59**, 1–32.

Vestergaard, K. (1982) Dust bathing in domestic fowls; diurnal rhythms and dust deprivation. *Appl. Anim. Ethol.*, **8**, 487–95.

Wilson, E.O. (1975) *Sociobiology. The new synthesis*, Harvard University Press, Cambridge, Mass.

Wood-Gush, D.G.M. (1973) Animal welfare in modern agriculture. *Br. Vet J.*, **129**, 167–74.

Part One

Manipulating Behaviour

2

Foraging behaviour

PETER W. GREIG-SMITH

All animals have to eat. The multitude of means by which they satisfy this need has always figured strongly in animal behaviour research, but has reached particular prominence in recent years, being at the forefront of the vogue for interpreting behaviour according to 'optimization' principles. No current discussion of the evolution and function of behaviour can escape reference to 'Optimal Foraging Theory'. The principles behind this approach, and the nature of the models generated, are well reviewed by Stephens and Krebs (1986). Such thinking has revolutionized opinions about foraging, at all levels from those of grand evolutionary design (Why is species A a herbivore, not a carnivore?) to the fine details of individual feeding schedules (How should a bird change its speed and direction of travel while searching for prey?). Though not without its critics (Pierce and Ollason, 1987; Gray, 1987), the optimal foraging ethic has undoubtedly advanced our understanding of behaviour in giant leaps, and in particular has stimulated new standards of experimental rigour and ingenious design in efforts to test the predictions of theory.

There has been a curious reluctance to apply the ideas stemming from optimal foraging theory to the solution of applied problems. This may be no more than a natural lag in the development of science from the first sparks of a new theory to the emergence of tried and tested practical applications. The speed of foraging research has indeed been dramatically fast, to the extent that the flow of published theoretical models has outstripped the direct tests of their predictions (Krebs *et al.*, 1983), let alone their conversion into innovative practical tools for animal management. There is, too, a difference between the type of practical experiments needed to bolster the development of a growing theory, and those required to further its applied uses. The first kind of experiment concentrates on achieving the minimum amount of replication needed to demonstrate the existence of predicted trends, while controlling for confounding variation such as habitat differences and changes in weather. In contrast, the second type demands evaluation of the robustness of effects under a wide variety of conditions, explicitly accepting the constraints which may limit or modify the underlying tendencies. In addition, the confidence in experimental results in applied research is generally higher than the academic researcher demands, thus often requiring more extensive testing with larger numbers of animals. All these considerations limit the direct use of results from tests of optimal foraging theory in applied problems, without an explicit investment in new research and development of a rather different kind.

The time is ripe to consider an increase in the exploitation of recent ideas about foraging behaviour, to allow manipulation of behaviour in a number of practical contexts (Metcalfe and Monaghan, 1987). There are many aspects in which animals' decisions may be open to manipulation. These include:

1. choices of where to eat, which are generally addressed by models of optimal 'patch use';
2. options of what foods to eat (prey selection models);
3. how long to feed at a particular place;
4. trade-offs between foraging effort and other activities such as anti-predator vigilance;
5. sensitivity of foraging behaviour to variability in feeding opportunities ('risk-sensitive' foraging models);
6. the particular issues connected with collecting and transporting food to a fixed centre of activity such as a nest ('central-place' foraging);
7. the utility of collecting and storing food ('hoarding') rather than feeding immediately.

These issues have all been addressed by recent optimization theory, but there are other aspects dealing with mechanisms that were debated much earlier, and which may be equally relevant to practical problems. These include 'search image' formation (Guilford and Dawkins, 1987) and conditioned aversion learning (Reidinger and Mason, 1983).

Although all these topics relate directly to foraging behaviour, it is neither wise nor possible to separate them entirely from other activities. For example, social behaviour in certain species is inseparable from individual foraging behaviour, because the allocation of time to feeding and vigilance depends on flock size (Caraco, 1979), or the presence of companions may compromise foraging prospects due to kleptoparasitism and other sorts of interference (Barnard and Thompson, 1985; Goss-Custard, 1980).

In this chapter I will explore a range of examples in which it may be

fruitful to apply ideas from 'optimal foraging' or other theory to practical problems of animal management — including some cases in
which this approach has already been attempted. First, however, it is
appropriate to consider *why* we might wish to manipulate foraging
behaviour, and some general principles about *how* it can be tackled.
The four major contexts in which this arises are pest control, wildlife
protection, food production and recreation. Finally, I will briefly
consider whether efforts to solve applied problems are likely to contribute useful feedback to the future development of theoretical ideas
about foraging.

2.1 WHY MANIPULATE FORAGING BEHAVIOUR?

There is a long history of humans using animal feeding behaviour to
their advantage. Some of this is direct interference, causing the animal
to do something in a different way to its normal activity (for
example, the taming and training of dogs for hunting). Other
relationships involve learning from an animal's behaviour without
causing it to change; this may or may not be detrimental to the
forager, e.g. in the use of seabirds as clues to profitable fishing sites
(Furness and Monaghan, 1986).

Not all of these mechanisms fall clearly under the rubric of 'manipulation', which must encompass the notion of *changing* behaviour to
human advantage, though not necessarily to the animal's own detriment. Indeed, such a wide range is covered that examples of almost
any combination of benefit and disadvantage can be found. A useful
way to categorize this diversity is to recognize that in any feeding
episode there are two partners — the feeder and its food. Either of
these may be advantaged or disadvantaged by human manipulation
(Table 2.1), forming four general strategies in manipulation of foraging
behaviour, with the feeder or its food as the ultimate 'target' for
change. This view helps to bring a unity to the wealth of different
examples, ranging from grazing to carnivory, from the simpler invertebrates to the higher vertebrates.

2.2 HOW CAN FORAGING BE MANIPULATED?

A useful starting point is to seek guidance from some intriguing
examples of 'natural' manipulation, which suggest the potential scope
and limitations of human intervention.

Animals that live in flocks or herds are particularly prone to influ-

Table 2.1 General scheme for the classification of cases in which foraging behaviour is manipulated. There are four broad objectives for manipulation, depending on whether the forager or its food is targeted, and whether the result is of advantage or disadvantage to the forager.

		Outcome of manipulation	
		Advantage	Disadvantage
	Forager	Encouragement to forager (e.g. ranching of game)	Interference with foraging efficiency (e.g. control of vermin)
Target of manipulation			
	Food	Protection of food (e.g. deterrence of crop pests)	Escalation of foraging (e.g. biological control of insect pests by predators)

encing each others' foraging. Some effects may be unwelcome, including kleptoparasitism by birds such as sparrows (which rob members of their own species that find profitable items of food or feeding sites) and others including gulls that specialize in an inter-species version of the same tactic (Brockmann and Barnard, 1979). Barnard and Thompson (1985) have shown that the feeding behaviour of lapwings (*Vanellus vanellus*) and the Golden Plover (*Pluvialis apricaria*) on arable fields is heavily affected by the attention of Black-headed Gulls (*Larus ridibundus*) which attack them to steal earth-worms. This research illustrates that animals are well able to specialize in a manipulative behavioural tactic when it is profitable, whereas the victims can make subtle changes in their behaviour to reduce the impact — in this case, plovers in mixed flocks tend to move away from gulls, but also modify their prey size selection to diminish the profitability, and hence the risks, of stealing by gulls.

Other interactions in feeding flocks may be more co-operative. Thus in many species there are characteristic call-notes or other signals that are used to indicate the location of a food supply to others. Even here, however, behaviour may be tuned to promote the benefits for certain individuals. Elgar (1986) has demonstrated that foraging house sparrows (*Passer domesticus*) that find a hidden supply of food only give calls to attract their fellows if the food is plentiful enough for the group (Figure 2.1). This is a subtle trade-off for the discoverer between ensuring adequate food and having the safety from predators that comes from feeding in a group.

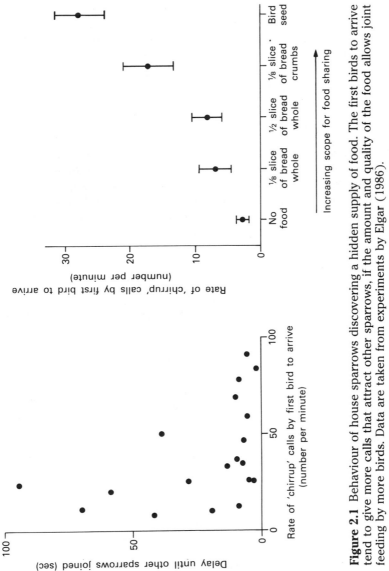

Figure 2.1 Behaviour of house sparrows discovering a hidden supply of food. The first birds to arrive tend to give more calls that attract other sparrows, if the amount and quality of the food allows joint feeding by more birds. Data are taken from experiments by Elgar (1986).

More widespread is the habit of many colonially-nesting or roosting birds of following successful foragers to profitable feeding sites on dispersal from their colony of roost. Although extensive research has failed to establish that this effect is the reason for the evolution of communal habits, it seems clear that at least some individuals benefit in this way (Brown, 1986; Greene, 1987). Generally, such behaviour is good for the follower, but may not be for the one who is followed, which has to bear an increased level of competition or interference. However, in inter-species associations there is more scope for mutually-beneficial sharing of food resources, by combining complementary, but non-competitive foraging roles. The best-known example of this kind is probably the African Greater Honeyguide (*Indicator indicator*) which uses vigorous displays to attract honey-badgers (*Mellivora capensis*) or humans to break open bees' nests, allowing the bird to feed on the larvae (Friedmann, 1955). The animals that do this for the bird also benefit, by gaining access to a supply of honey. Less direct, but equally important effects are founded on the fact that feeding by one animal prepares the way for others to feed more efficiently. For example, in African grassland, zebras' grazing removes taller-stemmed grasses, permitting wildebeest to feed on other grasses, followed by Thompson's gazelles, which take advantage of the protein-rich herbs made available, and the nutritious regrowth of the grasses eaten by the other species (McNaughton, 1979).

The advantage to one partner in mixed-species associations may be balanced by a counter-benefit of a different kind. Rasa (1984) studied an example of mutualism between dwarf mongooses (*Helogale undulata*) and hornbills (*Tockus deckeni* and *T. flavirostris*) in a Kenyan desert. The birds associate with foraging bands of mongooses, gaining benefit from the supply of food flushed by the mongooses. In return, the mongooses rely on the birds to provide early warning of predators, and reduce the time spent in vigilant 'guarding' by one or more of their number. This relationship is evidently important to both parties, since hornbills apparently attempt to 'chivvy' mongooses to start foraging, while the mammals will not leave their sleeping area until hornbills have joined them.

This type of co-operation is open to a sinister form of manipulation, in which predators may take advantage of responses to danger signals. For example, tropical forest falcons (*Micrastur* spp.) in Panama simulate the alarm calls of their small bird prey species in an apparent attempt to flush victims into the open, where capture is easier (Smith, 1969). Perhaps the most extreme of all 'natural' manipulation of feeding is that of nestling cuckoos and other brood parasites, causing their foster-parents to overwork in provisioning them, through exploitation of the potential of their 'super stimulus' begging to elicit an excessive response.

These few examples illustrate a useful general distinction between manipulation that is based on genuine changes in a foraging animal's options, to which it responds in an appropriate adaptive way, and cases in which such changes are mimicked, causing the animal to alter its behaviour unnecessarily. This contrast between 'honest coercion' and 'deception' underlies all forms of human intervention. The possibilities for deception depend on the flexibility of behaviour of the 'target' animal, and thus are likely to be greater for invertebrates such as insects, that can be controlled by fixed responses to odour cues, for example, than for the higher vertebrates, which are likely to learn to overcome such mistakes. Generally, deceptive manipulation will be less robust a tool than honest changes, but nevertheless may be more cost-effective in practice.

Lessons can be learned from experience in understanding the evolution of mimetic colouration in predator-vulnerable animals such as butterflies (Brower, 1984). This indicates that the risks of predation on a palatable mimic depend not only on its closeness of resemblance to a noxious model species, but also on the relative frequency with which predators encounter the two forms. Thus, infrequent judicious application of a deceptive tactic will help to reduce the risks of animals overcoming it by learning to identify and ignore the deception. In practice, there are three levels at which manipulation can be aimed:

1. changes to the habitat (e.g. planting pest-resistant varieties of crops to reduce damage, or altering the spatial distribution of an animal's food);
2. changes to the food (e.g. altering nutritional quality or food selection cues, or adding a repellant);
3. changes to the animal itself (e.g. affecting its experience of foods, or its hormonal balance).

Examples of these types are presented in the following sections. A further general point is that manipulation may be qualitative or quantitative. Depending on the circumstances, it may sometimes be easier to stop an animal entirely from using a feeding site than to influence its pattern of feeding while it is there. In other cases, the reverse may be true, and certain applied problems can be as well served by minor quantitative adjustment as by major changes in feeding habits.

2.3   PEST CONTROL

2.3.1 Habitat changes

Perhaps the most widespread need for foraging manipulation is in pest control, ranging from effective use of pesticides as a way of reducing pest numbers, to subtle deception of pests by influencing the predation risks under which they forage. The simplest level at which behavioural manipulation is invoked comes with the use of barrier crop protection. Total netting effectively prevents access by birds to fruit-trees, for example, but even an incomplete barrier may reduce damage. McKillop and Sibly (1988) review the behaviour of animals encountering electric fences, which are effective deterrents for some species.

Birds of many species are reluctant to forage below a network of fine wires stretched over a crop, apparently not only because these prevent them getting to their food, but also because the wires interfere with escape when alarmed. This method has been explored in the context of damage by herons (*Ardea cinerea*) and cormorants (*Phalacrocorax carbo*) at commercial fish farms. Draulans (1987) reviews the results of practical trials and experiments aimed at determining the optimal spacing for such impediments, which is apparently a close spacing (less than 1m apart) that interferes with flight even after birds have learned to avoid wider arrangements. This method of deterrence relies on a balance between the need to feed and the inconvenience or risks of physical access. That balance is liable to be disrupted when birds are short of alternative feeding sites, in which case other measures (e.g. screening the pond banks, or altering water depth) may be more effective. One intriguing solution is to use water jets to disturb the water surface of ponds, interfering with herons' ability to see the fish (A. Britton, personal communication).

Another approach to habitat alteration involves removing the refuges from which pests colonize crops. This depends on having detailed knowledge of the mechanisms and routes of immigration. In arable farming, there is concern that adjacent non-farmed habitats are a source of pests and weeds that colonize fields from their margins. The consequences of certain plant species (especially grasses such as *Bromus sterilis*) occurring in hedgerows around cereal fields are complex; they may spread as weeds, but also form a reservoir of barley yellow dwarf virus that is spread by aphids to the cereal plants (Marshall and Smith, 1987). This implies that problems might be alleviated by removal of hedgerows containing the host plants of pest species. Although that seems logical as a means of controlling single species of pests, there are many counter-advantages of retaining field margin habitats. For example, some species of invertebrate predators that prey on pests use field margins as winter refuges from which fields are colonized in spring (e.g. carabid beetles; Sotherton, 1985). The value of encouraging such natural predators may greatly outweigh the direct benefits of eliminating habitats that harbour pests, to the extent that the need for pesticide use may also be reduced (Burn, 1988).

In a slightly different way, immigration by weeds and ground-living pests can be usefully hindered by the creation of uncropped zones around the field margins to divert their foraging excursions. There is strong current interest in research to identify the most effective forms of management for these barriers — should they be cleared of vegetation by herbicides, or by physical tillage, and how wide should strips be? (Way and Greig-Smith, 1987; Park, 1988).

With bird pests also, crop damage often spreads from vantage points where birds gather before entering a field. Salvi (1987) has shown that starlings causing damage to vineyards in France are concentrated around pylons in mid-field, from which they fly out to feed. It may not be practical to remove these features, but bird-proofing of the critical vantage points may effect a substantial change in the distribution of feeding, which may be enough to eliminate economically-significant damage.

These examples are all gross changes to the habitat which foraging animals can detect by observation, or trial and error. However, their assessment of the likely profitability of feeding at a site can be more subtle. A good instance of a successful attempt to manipulate the foraging behaviour of a bird pest is that of the Brent Goose (*Branta bernicla*) in winter cereal fields in East Anglia (Inglis and Isaacson, 1978).

As flocks of Brent Geese approach, they circle over fields before landing, and their decision to land is influenced by the numbers of geese already there; the more birds present, the better feeding conditions are likely to be. However, this depends on what those geese are doing: if they are in their characteristic 'head-up' posture, indicating alarm, passing birds are less likely to join them that if they are busily feeding with heads lowered. Inglis and Isaacson (1978) found that by placing life-size plastic models of geese in fields, with various proportions of 'alert' and 'feeding' birds, they could influence the chances of new flocks landing (Figure 2.2). Geese tend to avoid fields with many alert models in favour of alternative areas where feeding models are used. The numbers of models should also have an effect, since although a high density should indicate a good site, competition there

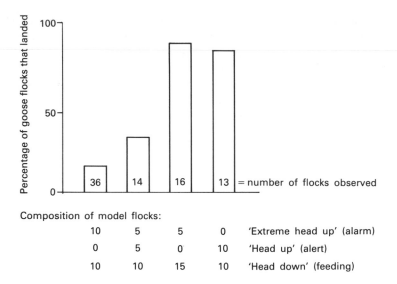

Figure 2.2 Numbers and percentages of flocks of Brent Geese that landed in pasture fields containing plastic models of geese feeding, or in alarm or extreme alarm postures. Data are from Inglis and Isaacson (1978).

may be correspondingly higher. Recent theory suggests that animals might take this into account and distribute themselves among sites according to an 'ideal-free' distribution in which the value of feeding depends both on food density and on competition (Sutherland and Parker, 1985). Thus there is an elegant management tool available to divert Brent Geese away from vulnerable fields into less important 'refuge' areas, using a 'deceptive' manipulation that can be finely tuned to affect the birds' behaviour quantitatively. The same principles might be used to manage other species, such as herons, which are known to rely on the presence of other herons when choosing feeding sites (Krebs, 1974). However, even when a system such as this works effectively, practical management depends on financial cost-effectiveness, and simpler, less sympathetic measures such as shooting and scaring may be expedient.

Another example of simple, but effective habitat change is the modification of grass-cutting practices on airfields, to discourage foraging birds, notably gulls and lapwings, that pose a hazard of air-strikes (Blokpoel, 1976). The traditional airfield management practice of having grass adjacent to runways cut to approximately 5cm provides good feeding conditions for lapwings searching for worms. A slightly longer sward of 15–20cm reduces the profitability of searching, and the birds are more likely to feed elsewhere (Brough and Bridgman, 1980). Studies of other species suggest that this empirical result may be due to a combination of short-term reduction in prey availability, difficulty in movement through longer grass, and a difference in prey selection between long and short swards (Eiserer, 1980).

In some situations, an incomplete, quantitative shift in foraging pressure may be all that is required to achieve adequate crop protection. One such example is provided by the habits of bullfinches (*Pyrrhula pyrrhula*), which damage orchard fruit crops by feeding on developing flower buds in the winter and early spring. The birds make forays out from adjacent woods or scrub, so that the accumulation of damage tends to be high at the edge of orchards, and low or absent in the centre (Greig-Smith and Wilson, 1984). Because fruit-trees are able to compensate for a surprisingly heavy loss of buds, resulting in no loss of yield, there is a relatively high threshold below which bullfinch feeding is economically unimportant. Therefore, instead of attempting to prevent bullfinches feeding in orchards — which is practically difficult — the same reduction of effective damage may be achieved by manipulating their behaviour so that feeding is distributed more evenly across the orchard, and damage to all trees falls below the critical threshold.

In an attempt to discover how this might be done, aviary experiments have been carried out, simulating the situation in which bullfinches fly from a refuge (a thicket of cut branches) to a series of feeding places spaced at various distances from cover (Greig-Smith, 1987). Among other results, this showed that a wider gap at the edge of the 'orchard' encouraged bullfinches to spread their feeding more evenly, but a buffer zone of less-preferred food did not (Figure 2.3). These findings are qualitatively consistent with an optimization model of foraging behaviour based on the principles of 'central-place foraging' (Orians and Pearson, 1979) which predicts the best strategy for animals to harvest food (i.e. how far to travel, how much food to collect, etc.) when it has to be carried back to a 'central place', such as a nest site. In this case, the energetic costs of travel over progressively longer distances affect the value of feeding there, and the amounts of food collected. This approach could be developed further, to predict behavioural management of many situations in which pest damage spreads unevenly from non-agricultural land.

The costs of travel are not merely energetic, however; the risk of predation is also likely to increase at greater distances from cover. Several recent studies have demonstrated that animals such as birds (Lima *et al.*, 1987) and squirrels (Lima *et al.*, 1985) apparently take this factor into account, alongside foraging profitability, in organizing

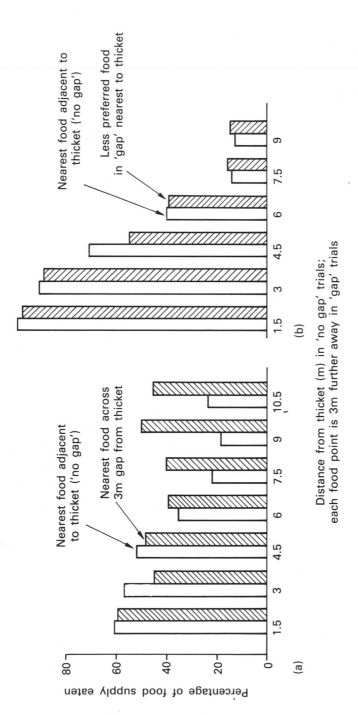

Figure 2.3 Food consumption by bullfinches at various distances from cover (a thicket of branches in a large aviary), in relation to the width of the gap to be crossed to reach the food. The birds' daily food consumption was similar in both conditions. However, their foraging effort was spread farther by a wider gap (a), but not if the gap contained a less favoured type of food (b). Data are from Greig-Smith (1987).

their foraging trips. The practical implication is that the density of vegetation in the habitats surrounding fields, and field size, are likely to influence the behaviour of pests, and perhaps the levels of damage they cause.

2.3.2 Changes to food

The desired reduction in pest damage may come about from altering the food itself. In animal feedstuffs, bird depredation may be dramatically affected by a physical change in texture or particle size. For example, Feare and Wadsworth (1981) found that the feeding of starlings from troughs of 'complete diet' cattle food (i.e. a mixture of all required nutrients) could be alleviated by changing the preferred cereal grain component from rolled barley to ground barley or brewers' grains. These smaller particles are more difficult for starlings to separate from the mixture and handle rapidly, and their feeding profitability declines.

Pests can be effectively diverted by provision of a preferred, alternative food. For example, Sullivan and Sullivan (1982) showed that predation of lodgepole pine seed by deer mice (*Peromyscus maniculatus*) could be greatly reduced by mixing the sown confer seed with sunflower seeds, which are preferred by the small mammals (Figure 2.4).

For growing crops, it is theoretically possible to make use of the fact that animals often have distinct preferences for certain cultivated varieties over others. In the battle against invertebrate pests, it is common to select cultivars on the basis of resistance to particular insect pests, both in the course of breeding programmes and in the farmer's choice of seed from various alternatives, if pest pressure varies locally. Similar protection can be afforded by the use of pesticides, which may avoid the compromise over reduced yield *vs* improved resistance that can be achieved by choosing a resistant cultivar, although there is reason to minimize pesticide use, for environmental protection. With vertebrate pests, the position is less clear: the basis of cultivar resistance is less well established; the economic pressure to eliminate damage is generally less than for invertebrate pests; and there are fewer proven options for chemical control. Nevertheless, there are marked preferences among varieties in many bird pests (Table 2.2) and levels of bird damage are high enough to warrant the development of rating systems for bird-resistance (e.g. in maize, Dolbeer *et al.*, 1982; and in sorghum, McMillian *et al.*, 1972). Preferences may be based on physical features that affect the profitability of foraging through handling difficulties

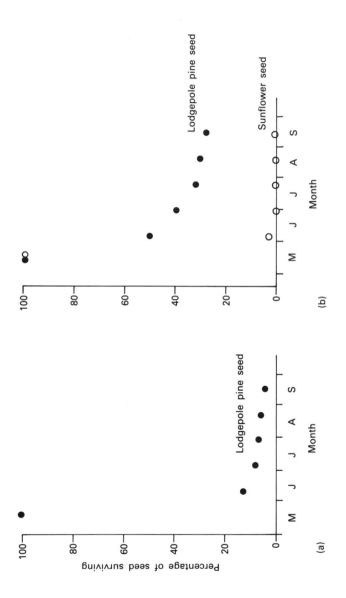

Figure 2.4 Differences in the seasonal pattern of predation by small mammals on lodgepole pine seed, in experimental plots with or without additional supplies of sunflower seed, which was preferentially eaten, reducing the loss of pine seed in high rodent density areas. (a) lodgepole pine seed sown alone; (b) sown with sunflower seed. Data are from Sullivan and Sullivan (1982).

Table 2.2 Feeding preferences of pest birds among cultivated varieties of crop plants

Crop	Bird species	Features correlated with preference	Reference
Sorghum — 142 varieties	Red-winged blackbirds (*Agelaius phoeniceus*) and other species	Tannin content, seed colour, plant height	McMillian *et al.*, 1972; Bullard *et al.*, 1981
Cherries — 12 varieties	Starlings (*Sturnus vulgaris*)	Fruit colour	Stevens and De Bont, 1980
Pears — 36 varieties	Bullfinches (*Pyrrhula pyrrhula*)	Bud size, volume of budscales, simple phenolic chemicals	Summers and Huson 1984; Greig-Smith and Wilson, unpubl
Medics — 7 species (*Medicago* spp.)	Skylarks (*Alauda arvensis*)	Not determined	Halse and Trevenen, 1985
Sunflower — 17 genotypes	Red-winged blackbirds	Polyphenol content, flat flowerheads, tightness of seed packing	Fox and Linz, 1984
Maize — 10 hybrids	Red-winged blackbirds	Soluble tannins, energy content	Mason *et al*, 1984

(Dolbeer *et al.*, 1982), or on chemical characteristics affecting digestibility (Bullard *et al.*, 1980).

Further development of this approach demands a new application of theory to help explain how physical and chemical features are weighed against each other in birds' preferences. Optimization models of diet choice have been good at defining how animals assess *quantities* of food and the behavioural costs of handling food in their choices among alternative options. There has been less progress regarding the *quality* of food, which has to be assessed more realistically than by its 'energy content' (Robbins, 1983; Belovsky, 1984). Most animal foods consist of a range of diverse nutrients that are hard to convert into a single 'currency' comparable to the 'energy/handling time' ratio used in the usual diet choice models (Stephens and Krebs, 1986). In addition, foods may contain adverse chemical components (e.g. toxins), to which animals may react in a qualitative fashion that can outweigh their quantitative preference for the nutrient component (Rhoades and Cates, 1976).

The presence of deleterious secondary chemicals in feeding preferences raises another possibility for manipulating behaviour, by application of formulated chemical repellents to crops. There have been many efforts to develop effective repellents, for use against a wide range of species (particularly birds) on many crops around the world (Wright, 1980). These can be regarded as working through one of three mechanisms:

1. inherent aversion to substances with distasteful characteristics;
2. learned aversion to chemicals that cause illness after ingestion;
3. avoidance of novel substances.

In reality, all three mechanisms can work together in some cases — a novel agent of a learned aversion may also have a repellent flavour. The most successful deterrent in recent years has been the pesticide methiocarb, which is regarded as operating by conditioned aversion learning (Rogers, 1978), and has shown evidence of reducing bird damage to a wide range of crops (Benjamini, 1980; Conover, 1985). However, the use of this type of substance is not entirely benign, for in order to be effective, a conditioned aversion agent must be sufficiently toxic to cause at least mild physiological distress, and there is often a narrow margin of safety between the repellent dose and the lethal dose (Schafer *et al.*, 1983). In addition, there may be a risk of mortality of more susceptible non-target species.

One way of minimizing these problems is to pay attention to the sensory modalities and other mechanisms by which animals chiefly select their food (Table 2.3). For birds, this generally means visual cues rather than flavours or colours, though the reverse may be true for most mammals. Several researchers have explored the utility of adding conspicuous colours to repellent formulations in order to enhance birds' aversive responses (Table 2.3). This may allow the amount of the toxic ingredient to be reduced to a level below that which would be effective in an uncoloured treatment (Elmahdi *et al.*, 1985; Figure 2.5). In addition, some contexts might allow the active ingredients of repellents to be minimized by treating only part of the crop (e.g. a vulnerable headland) with a coloured formulation, and using the colour alone over the remainder, relying on birds to extrapolate from the aversive portion they have sampled, and avoid a whole field. Partial treatments may cause disproportionate aversion (Figure 2.6), in aviary experiments but on a practical scale the deception depends on the balance between the birds' contact with treated and untreated parts of the crop, which may be difficult to control.

Table 2.3 Aspects of chemical repellency in pest birds illustrated by methiocarb treatments on crops

	Species	*Effect*	*Reference*
1. Immediate distaste	House finches (*Carpodacus mexicanus*)	Choice tests indicate that birds rely on visual cues rather than taste	Avery, 1984
2. Rapid physiological response	Red-winged blackbirds (*Agelaius phoeniceus*)	Increase in heart rate and respiration rate, 5-min after dosing	Thompson, *et al.*, 1981
3. Avoidance of novelty	Grey-backed white eyes (*Zosterops lateralis*)	Naive birds were more easily deterred from feeding on grapes than experienced birds	Rooke, 1984
4. Conditioned aversion learning	Red-winged blackbirds (*Agelaius phoeniceus*)	Two 5-min exposures produced aversion lasting 16 weeks	Rogers, 1978
		Aversions are to the food if repellent is hard to detect but to the chemical if it is detectable	Conover, 1984
5. Added colour or flavour cues	Red-billed Quelea (*Quelea quelea*)	Addition of wattle tannin (astringent) or calcium carbonate (white) allowed reduction in quantity of methiocarb used	Elmahdi *et al.*, 1985; Bullard *et al.*, 1983
	Red-winged blackbirds (*Agelaius phoeniceus*)	Birds generalize aversions to shades of red, but not green cues	Mason and Reidinger, 1983a
6. Social learning	Red-winged blackbirds (*Agelaius phoeniceus*)	Birds avoid treated food after observing signs of distress in others	Mason and Reidinger, 1982

Table 2.3 continued

	Species	Effect	Reference
	Starlings (*Sturnus vulgaris*)	Reluctance to associate with individuals present when birds felt ill	Mason and Reidinger, 1983b
7. Partial food treatments	House finches (*Carpodacus mexicanus*)	Birds avoided both treated and untreated seeds when presented as a mixture	Avery, 1985

Field tests have produced equivocal results (Conover, 1985; Bailey and Smith, 1979).

An alternative approach to the use of potentially toxic synthetic pesticides is to explore the aversive potential of naturally-occurring plant and animal products. The presence of highly distasteful secretions in aposematically-coloured insects is well known (Edmunds, 1974), and some of these might be used as deterrent treatments, although their effects apparently often depend on a complex mixture of chemicals which might be difficult to synthesize, rather than a single active principal. The same drawback applies with even more force to odours of mammalian predators, which are not only difficult to reproduce, but are impressively variable among individuals, although their repellent effect has been demonstrated experimentally (Sullivan *et al.*, 1985).

Plant chemistry may offer more hope of success. Feeding preferences in the wild have been shown to correlate with levels of phenolic components, for example, in animals as diverse as detritivores (Valiela and Rietsma, 1984), insect herbivores (Tahvanainen *et al.*, 1985), birds (Buchsbaum *et al.*, 1984) and mammals (Cooper and Owen-Smith, 1985). In many instances, experimental application of extracts from plants, or of the individual chemical components, have deterred animals from treated areas (Bryant, 1981; Buchsbaum *et al.*, 1984). However, as with all repellents, their effect declines when the animals are short of an alternative on which to feed. Perhaps repellency is an approach which can only provide a useful tool under limited, favourable conditions.

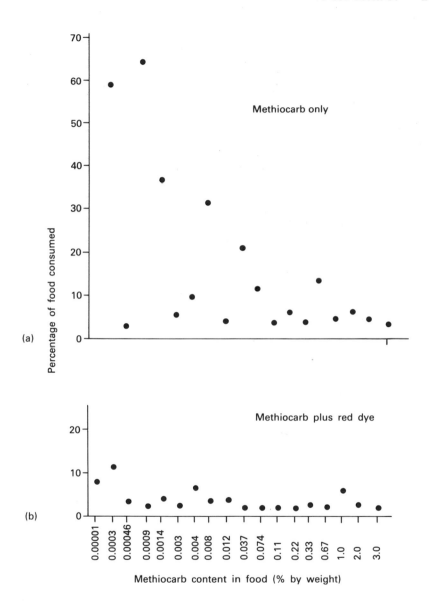

Figure 2.5 Effect of adding a colour cue to a bird repellent chemical. In laboratory experiments, starlings ate much less red-dyed food (b) treated with low levels of methiocarb than if the food was uncoloured (a). Points represent average consumption by individual starlings during short feeding trials on five successive days (Greig-Smith and Webb, unpublished data).

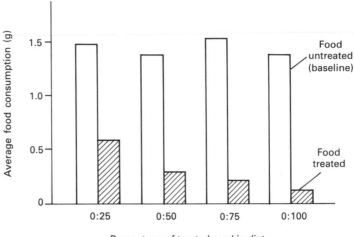

Figure 2.6 Levels of feeding by captive pairs of house finches (*Carpodacus mexicanus*) on safflower seed treated with 0.2% methiocarb, according to the ratio of treated to untreated seed in the food. Columns represent average consumption in short trials on five consecutive days, with (shaded) and without (unshaded) treated seeds in the diet. Data are from Avery (1985).

2.3.3 Altering the animal or its social environment

Foraging decisions are mediated through social effects in many species. On the one hand, a food supply that is profitable to an individual may not be worth exploiting if the animal is in a competitive group, particularly if the group includes social dominants. Feare and Inglis (1979) found that restricting the amount of space available to a flock of feeding starlings was sufficient to reduce their efficiency. This might apply to measures for restricting birds' access to cattle feeding troughs through quite minor changes to their design.

Feral pigeons in towns and cities cause problems by fouling buildings and food supplies where they feed on spilt grain. A major element is their social habit, birds often gathering in flocks of several hundred to feed (Murton *et al.*, 1972). Population control by killing or the use of chemosterilants is not feasible (Feare, 1986), but recent research on foraging strategies has provided greater understanding of their behaviour that may help in alleviating the problem. Feral pigeons rely on exploiting a very variable, unpredictable series of food sources, at which their feeding efficiency depends heavily on competition from other pigeons. For example, Lefebvre (1983) has shown that in stable conditions pigeons learn to distribute themselves among the available sites in an appropriate equilibrium depending on the relative profitability (food eaten per minute) of feeding at different sites, but initially their combined exploitation is inefficient. To reduce their performance, it might therefore be sufficient to 'confuse' birds by ensuring a variable and unpredictable distribution of food sources. Paradoxically, this could involve provision of additional food, which might profitably be combined with baiting to divert birds away from areas of greatest nuisance, rather than to attempt control by the licensed use of narcotic or toxic baits.

Aversion learning can be a useful mechanism in the pest control context, but it may sometimes create problems. The control of rodent pests relies largely on the use of poisons, notably anticoagulant rodenticides, placed in food baits where rats and mice are likely to ingest a lethal dose. This may be disrupted if the animals develop an aversion to the flavour of the poison before they have eaten enough, or if they display 'bait shyness' due to reluctance to feed on a novel food (Robbins, 1981). Several ways of circumventing these difficulties have been tried. Mason and Reidinger (1983a) and Stewart *et al.*, (1983) have explored the taste qualities of rodenticides in relation to the development of conditioned aversions. This raises several interesting possibilities for improving control. Similarity of taste between a rodenticide and harmless chemicals (e.g. between strychnine and sucrose octacetate) raises the possibility of pre-baiting with an agent that resembles the poison, so that rodents find it difficult to detect a change when the toxic bait is introduced.

Another strategy discussed by Mason and Reidinger (1983a) is to use taste resemblance to deter rodents from crops. If the animals are conditioned to avoid a flavour by association with a toxic bait placed near to a crop, the crop itself should then be protected by application of a harmless, similar-tasting chemical. Finally, they considered whether bait-shyness could be overcome through making use of rats' grooming behaviour. Application of aversive chemicals to the fur (as might be achieved by use of commercial rodenticide dusts and powders) led to more ingestion than if the same substance was offered in food or water. In studies with pine voles (*Microtus pinetorum*), grooming an aversive agent from the fur was associated with elevated levels of blood corticosteroids, which are believed to stimulate further grooming, and hence ingestion of more of the chemical. This stereotyped lethal feedback loop could be a potent means of enhancing the uptake of poison.

Manipulations for pest control do not always involve detriment to the foraging animals. In biological control programmes, predators (and parasites) of pests are used to reduce infestations, and reduce the need for chemical insecticide treatments. In cereal growing, for example, a broad range of polyphagous and specific invertebrate predators help to control field populations of aphids (Potts and Vickerman, 1974; Sunderland, 1975). Their impact can be enhanced in several ways: by deliberate introduction of captive-reared predators; by adjusting the timing of the crop husbandry (sowing date, timing of pesticide applications) to allow natural predators to have an impact; by promoting the dispersal of species that overwinter in field boundaries through sympathetic management of field margin habitats; and by reducing chemical inputs that may kill the beneficial species as well as the pests. Practical integrated pest management is much further advanced in horticulture, however, where predatory mites are an important means of control of pests (Payne, 1988).

Introduction or encouragement of predators may have effects other than mortality of pests. The need to compromise feeding efficiency when risk of predation is high may lead to a change in the foraging area exploited, in the division of time between feeding and anti-predator vigilance, and in feeding bout lengths (Milinski and Heller, 1978). Experiments to explore the implications of these changes were conducted by Werner *et al.* (1983), who introduced predatory large-mouth bass (*Micropterus salmoides*) into ponds containing fish of three size-classes of bluegill sunfish (*Lepomix macrochirus*). Only the smallest bluegills were vulnerable to direct predation and, in the presence of the bass, they altered their foraging behaviour to favour vegetated habitats, even though their food intake there was much lower than in open habitats. The larger fish continued to feed in the productive open areas. Because of their reduced intake, the small bluegills grew more slowly than in the absence of predators, whereas the larger bluegills grew more quickly, as competition for food in the open water was reduced. Effects of this kind open up possibilities for management of commercial or angling fish stocks, using a balance of habitat availability and predation pressure to control the growth and productivity of fish delicately. See Chapter 5 for a more detailed discussion of ways of improving productivity in farmed fish.

It may not be necessary actually to introduce a predator if its presence can be simulated. Birdscaring teams on airfields, for example, have tested a variety of predator models of varying realism; none has been as successful as the use, even occasionally, of a live tame hawk (Blokpoel, 1976). However, see Chapter 3 for further information on predator simulation.

In general, the current approaches to pest control are shifting progressively towards non-lethal methods that involve manipulation of behaviour, and away from mass killing, in recognition of the futility of attempts to eliminate a sufficiently large proportion of any pest population (Feare *et al.*, 1988).

2.4   WILDLIFE PROTECTION AND CONSERVATION

The principles of diet selection have been useful in the context of pest control; they are equally relevant to the protection of non-target species of wildlife from the adverse effects of crop protection measures. In particular, the mechanisms by which animals learn to avoid toxic chemicals in their food serve to protect them from harm through eating pesticide-treated seed (Greig-Smith, 1988a). Many insecticides are repellent to birds (Schafer *et al.*, 1983) but there is scope to develop improved formulations of such pesticide treatments to include more specific repellents. For example, woodpigeons (*Columba palumbus*) and related species seem to be unusually vulnerable to poisoning because they feed heavily on dressed seed without evidence of unpalatability. Study of the taste capabilities of pigeons to further the work of Duncan (1963) might reveal chemicals that would be the basis of effective synthetic repellents.

There is one major difference between the application of such effects in pest control and in wildlife protection. In one case, the aim is to reduce the combined total amount of feeding by animals within an area, whereas in the other, concern centres on the amounts of food taken by individuals, so as to limit their risks of accumulating a toxic dose. In wildlife protection contexts there is, therefore, more interest in individual variance of behaviour (e.g. to identify particularly vulnerable age-classes or social categories), and in the thresholds at which animals decide to feed or stop feeding. This means that optimization models of foraging are even more applicable than in pest control, since these models do not generally make predictions about combined feeding pressure within a population (but see Sutherland and Parker, 1985).

A different level of manipulation occurs in the context of wildlife reserves. Here, foraging behaviour is closely tied to population density, social structure and ranging behaviour, to the extent that manipulation may be through changes apparently remote from foraging, although their effect is eventually felt in a change of feeding behaviour.

The principal issue in establishment of nature reserves is selection of the size and shape of plots to be protected. Although this is generally constrained by many non-biological factors, it is useful to begin by

considering the theoretical pros and cons of various possible arrangements of land. Many authors have recently explored the implications
of 'island biogeography theory' (MacArthur and Wilson, 1967) for
questions such as whether a single large plot of protected land makes
a better reserve than many small ones (Margules *et al.*, 1982; Woolhouse, 1983). The answer to this question depends on whether the
purpose is to maximize the number of species retained in the habitat,
or to safeguard particular species. Thus, for example, a range of patch
sizes should encourage species with large territorial requirements as
well as those which favour edge habitats, whereas there may be a
single minimum size suitable to ensure viable local populations of an
endangered species. For this reason among others, there has been
little agreement about general rules for optimal design of nature
reserves.

Foraging fits into this issue through its influence on the extent of
animals' ranging behaviour, and its flexibility. Island biogeography
principles apply most readily to animals that forage within permanent
territories of a defined size; they are less useful in cases where
animals become nomadic or migrate outside their breeding season. In
those cases, the success of reserves may depend more on the provision and location of critical food resources (e.g. rare fruiting trees in
tropical forests) than on the size of the protected area. Animals may
themselves contribute to the suitability of their habitats. Many species
of trees bear fruits adapted for dispersal by birds, which can be instru-
mental in the local spread of important food plants, that may be
present as rare relics, or can be introduced if necessary (Howe, 1984).
Only detailed study of the feeding habits of the animals will provide
sufficient understanding to manipulate the availability of such
resources.

Where a reserve area is composed of a number of patches that are
relatively small compared with the ranges of animals of concern,
simple connection by habitat 'corridors' may be beneficial. At first
sight, this seems to be most valuable for poorly dispersive species, as
a means of preventing local extinction within isolated habitat patches.
However, a converse view is that corridors would be most heavily
used by dispersive animals, so that the chief outcome might be to
increase populations of these species by permitting further and more
frequent movements among alternative food sources. There is surprisingly little empirical information on the use of corridors, although
some evidence is cited by Baudry (1988) that the distribution of plant
species in hedgerows on agricultural land is influenced by their
'connectedness' to woods (presumably through seed dispersal,
reflecting the foraging movements of herbivores), and that the

presence of beetles in hedges depends on the distance from connected woods. Henderson *et al.* (1985) also found that chipmunks in Canada used fencerows to colonize patches of woodland. These considerations apply to animals which are reluctant to forage away from cover, but there are, of course, many that require large open areas as protection from surprise attack by predators (e.g. plovers on arable fields; Barnard and Thompson, 1985).

One obvious consequence of defining a protected nature reserve is the creation of a boundary that may not coincide with the natural limits of the animals' movements. The problems of overcrowding of large mammals (e.g. elephant, hippopotamus) in some African game parks are partly due to the migration of animals outside the park boundaries In response to overgrazing, with consequent risks of poaching and agricultural damage (Eltringham, 1979). This may be alleviated by short-term contingencies such as establishing barriers of less favoured habitats to inhibit movements, but really requires sustained management of population levels, which may or may not involve culling. For some species, it is remarkable how effective a man-made barrier may be. For example, carabid beetles and small rodent populations may be effectively isolated by their reluctance to cross roads (Figure 2.7), although other species (e.g. badgers, *Meles meles*) are so difficult to divert from traditional foraging routes that road-builders have resorted to construction of tunnels to allow the animals a safe passage.

Having established a wildlife reserve, it must be managed, and this involves considerable indirect manipulation of foraging behaviour. In the protection of 'natural' areas and their wildlife, operations for intervention are limited, essentially allowing either (1) habitat management or (2) control of population levels by culling and encouragement of

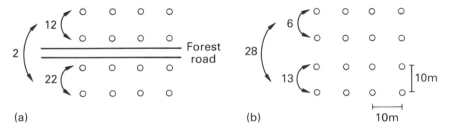

Figure 2.7 Effectiveness of roads as barriers to small mammals, illustrated by the numbers of yellow-necked mice (*Apodemus flavicollis*) moving between parallel lines of live-traps (a) across a forest road, and (b) in a uniform control area. Data are from Mader (1984).

reproductive success. Nevertheless, these measures have dramatic effects on foraging.

In African savanna and grassland areas, there is a diverse community of grazing and browsing herbivores, each with different dietary preferences, selectivity and food requirements (Eltringham, 1979). For some species, prior feeding by another animal is necessary to provide appropriate foraging conditions. One example involves the zebra, wildebeest and Thompson's gazelle mentioned above, and another is the removal of tussock grasses by buffaloes, allowing the development of short, spreading grasses preferred by hippopotamus and warthog (hippopotamus grazing encourages further growth of tussock grass, thereby completing a mutualistic cycle). Conversely, competition between animals (e.g. browsing antelopes) may help to hold populations in check, although the broad diets of most browsers permit some overlap before significant competition is felt. Artificial manipulation of the population of any one species is liable to affect the foraging prospects, and hence perhaps the population dynamics of others, illustrating the interdependence of species within the community.

Use of controlled burning to manipulate food supplies more directly is a powerful tool in African reserves. Periodic burning of small areas provides a mosaic of patches at different stages, appropriate to the foraging habits of various species. By changing the burning regime, it is possible to alter the balance of feeding opportunities, and hence the composition of the herbivore community, to favour chosen species or discourage others. The success of such measures depends heavily on animals' flexibility in diet and feeding techniques, which in some cases corresponds to the predictions of optimal-diet models (Owen-Smith and Novellie, 1982).

Knowledge of foraging behaviour can also be of value in programmes to re-introduce captive-bred animals of endangered species to areas where they have declined or disappeared. Several such projects have been attempted, with varying success (Temple, 1978). Often, a major cause of the declines of such species is the removal of a critical food supply in the course of agricultural cultivation, forestry, or other human activities. Despite the loss of typical habitat, animals may sometimes adapt to changing environments if they can be persuaded to alter their diets. Training of captive animals may help to achieve this when they are released in the wild, through a gradual progression from reliance on bait stations to full independence. Social learning of foraging skills and novel food preferences may then reinforce new habits for survival.

2.5 FOOD PRODUCTION

More is known of the feeding behaviour and physiology of domesticated animals than of any wild species. The management of farm
animals is discussed in more detail in Chapter 6, but it is appropriate
to outline some general points here. In particular, there is a difference
between the opportunities for manipulation of animals held on
enclosed fields and those that are free-ranging in more natural habitats. This stems from the aim of domestic stock management (and
ranching of animals such as deer and antelope), to maximize meat or
milk production. The concept of a balanced, self-sustainable population that is the cornerstone of much conservation management is
therefore irrelevant; in most cases the numbers and biomass of stock
are considerably higher than could be maintained by a free-living
population. The composition of animal groups is also generally less
variable, in sex, age and body condition, than in wild populations. All
these factors influence the options for manipulation.

To maximize productivity, there are four general possibilities for
manipulating foraging behaviour:

1. choosing *what* the animals are feeding on (e.g. by movement from
   field to field, rotation of crops, etc);
2. controlling *where* they feed (e.g. the use of strip-grazing to spread
   feeding pressure);
3. selecting the *numbers* and *mixture* of animals allowed to forage as
   a group (e.g. competition, social facilitation and complimentary
   feeding habits of different species can all be exploited as mechanisms to alter subtly the feeding pressure on sward);
4. limiting the *time* for which they can feed (e.g. to control the
   grazing pressure on grass in order to influence its future growth).

Animals such as sheep, cattle and horses respond to constraints of
this kind in many ways, that may involve changes in the duration of
feeding bouts, the rate of feeding, the amount taken per bite, and their
selectivity for plant species and plant parts providing preferred nutrition. Many of these aspects are summarized by Arnold and Dudzinski
(1978). The immediate ends of such manipulation may be directly to
improve the performance of the stock themselves, or to influence their
food supply, by controlling the changes in growth rate and nutritional
status that accompany different levels of grazing.

The practical strategies used to achieve these aims can be
developed through an understanding of the factors determining diet

selection. For example, Illius (1986) used a simulation model of feeding by grazing mammals, based on sward height, digestibility of vegetation and the size of the animal. The detailed quantitative predictions from this approach, once validated, offer a strong tool for designing grazing management programmes.

There are similarly wide possibilities for manipulation of the feeding behaviour of fish, as part of commercial fishery management. These are discussed in detail in Chapter 5.

In the context of crop production, most animal feeding behaviour is unwanted, but one exception is the role of pollinators. Among orchard fruits, in particular, farmers rely heavily on honeybees to effect a sufficiently high degree of pollination. The extent of manipulation is limited to decisions about when and where to place hives in the orchard, and to the choice of pollinator varieties to grow among the main crop. Knowledge of bees' sensory capabilities, the criteria they use in exploring flowers for pollen and nectar, and their preferences for the flowers of different fruit-tree varieties can all contribute towards refining this use of honeybees. It must be recognized, however, that wild insects (e.g. bumblebees) may be more effective pollinators in some crops, though their presence and numbers obviously cannot be managed in the same way, at least in field crops.

Manipulation is also required to safeguard bees by preventing them from foraging where they should not. The modern farming landscape contains many attractions, such as fields of flowering oilseed rape, which bees will travel long distances to exploit (Free and Ferguson, 1983). These crops are liable to receive periodic pesticide treatments, with consequent hazards to foraging bees. Application of the insecticide triazophos to control seed weevil on rape plants presents a particular risk, because the optimum timing for treatment overlaps with the period of attractiveness to bees. As a compromise, such treatments are permitted only after 90% of petals have fallen from the crop, which is later than the ideal timing for pest control (Greig-Smith, 1988b). Even then, bees may be killed if the flowering of the crop is patchy, and there may be scope to alleviate problems of this kind by more detailed interpretation of bees' foraging tactics. Much is known about the ecology and behaviour of honeybees (Seeley, 1985), including recent experiments that demonstrate their conformity to the predictions of optimization models concerned with patch use (Krebs *et al.*, 1983). Despite this, the practical manipulation of bees to promote pollination is largely restricted to movement of hives.

2.6   RECREATION

With the growth of tourism throughout the world, the commercial possibilities of wildlife spectacles have become increasingly obvious, leading to extensive management with tourist appeal in mind. Many of the principles discussed above in relation to species protection are equally applicable to management of reserves for the interest of visitors, and the two functions generally go hand-in-hand. However, more direct manipulation is practised in order to ensure that animals provide a good view of themselves, by deliberate placement of salt-licks, waterholes, bait stations and other attractions where they are best displayed to visitors.

One recent example of conservation achieved by means of a care-fully managed exercise to display wildlife to the public is described by Clark (1987). A mass pre-migratory roost of several species of swal-lows in Ontario, Canada provides an annual summer spectacle of many tens of thousands of birds engaged in aerial aerobatics at dawn and dusk. A concerted publicity campaign led to sufficient revenue from visitors to protect the site from development. Though the birds forage without direct manipulation, the preservation of the habitat at a single traditional site was a practical measure that is both the end and the means to an educational and conservation aim. Clark's account is also of interest in demonstrating the pragmatic use of economic analysis to quantify the tourist potential of natural phenomena, through the public's 'willingness to pay'.

A much earlier example of recreational manipulation is seen in the management of grouse moors and other habitats to promote game-bird populations (Hudson and Rands, 1987). The vegetation of grouse moors is dominated by ling heather (*Calluna vulgaris*), and a simple manipulation of this plant, by periodic burning, achieves a far-reaching effect on the populations of red grouse (*Lagopus scoticus*), mediated by the birds' feeding behaviour.

In spring, over 90% of the diet of breeding grouse consists of heather, and there is evidence that the birds select plant parts rich in nitrogen and phosphorus (Moss, 1972), with their subsequent breeding performance related to the levels of these nutrients. Since the chemical quality of heather varies with the plant's age, major changes in grouse breeding and population dynamics can be brought about by manipulating the availability of different age-classes of heather. Management of the moors to favour grouse populations consists chiefly of cyclical burning, to ensure a plentiful supply of young growth, which is both nutritious and digestible (Savory, 1978).

The interaction of game interests and modern agriculture has produced the need for another form of management to favour game-birds on farmland. Heavy declines of partridges (*Alectoris rufa* and *Perdix perdix*) in the south of England in recent decades have been attributed to the increasing use of pesticides on arable crops, particularly cereals (Potts, 1986). Insecticides have reduced invertebrate numbers directly, while herbicides have done so indirectly by eliminating many of their food plants. Combined with the loss of hedgerow nesting habitat, and close cultivation at the edge of fields, these trends have greatly reduced the suitability of farmland for game-birds. Recent studies have demonstrated that a compromise can be reached simply by not spraying herbicides on the outermost six metres of cereal fields; the resultant increase in herbs and insects at the field margin allows higher breeding success and survival for partridges and pheasants, while not reducing crop yields unacceptably (Rands, 1986; Boatman and Sotherton, 1988). (See Figure 2.8.)

2.7   CONCLUSION

This review has deliberately covered a wide range of examples in order to demonstrate the far-reaching potential of behavioural manipulation in solving practical problems. Most of the examples concern birds, for good reason. In part, the behaviour of domesticated mammals and of fish are considered in more detail in other chapters, but the chief reason is that tests of foraging theory have been dominated by studies of birds, and indeed, general theoretical predictions have often been developed with feeding birds in mind.

This introduces one area in which applied research involving foraging behaviour can feed back to influence the development of theory. With the luxury of free choice, researchers wishing to conduct experimental tests of theoretical predictions naturally tend to work with species whose characteristics match the assumptions of their models. For example, economic models of food intake have been tested with a variety of small insectivorous passerine birds (Krebs *et al.*, 1983), because these are assumed to be under strong pressure to maximize intake rates in winter foraging. Whatever the results of the experiments, the untested assumptions remain, and accumulating published studies on a similar type of species leads to a distorted view of the factors that may be of more general importance. Applied studies can help to redress this imbalance. With the constraint of having to investigate a particular problem, foraging theory can be applied only by facing up to the validity, or otherwise, of the underlying assumptions to the species in question. One example is in the criteria animals

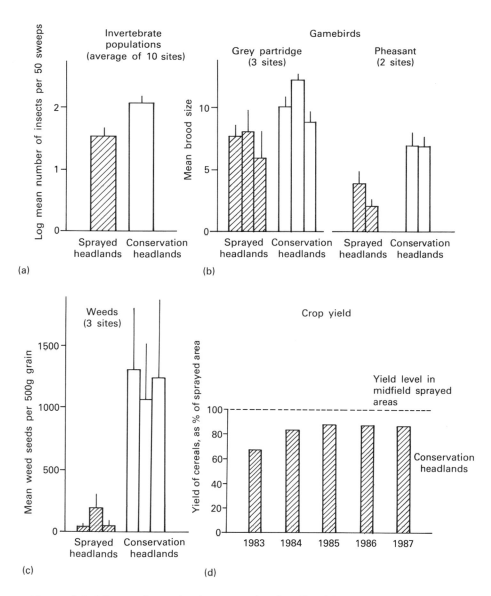

Figure 2.8 Effects of creating 'conservation headlands' around the margins of cereal fields, in which pesticide use is reduced in the outer 6 metres of the crop. (a) Density of invertebrates was increased in conservation headlands; (b) survival of gamebird chicks was enhanced; (c) weed populations, and numbers of weed seeds in the harvested grain, were greater; and (d) there was a slight loss of yield in the headland, compared with parts of fields that were fully sprayed. Data from Rands (1985, 1986) and Boatman and Sotherton (1988).

use when choosing prey. Foraging theory would suggest that the decision is an economic one and that view is bolstered by experiments in which palatable prey differing only in size and handling costs have been compared. However, as the examples quoted in this chapter illustrate, simple economics are often outweighed by reactions against unpalatable constituents, learned aversions, and other aspects that have yet to be fully incorporated into current optimization theory in the way that trade-offs with vigilance have been.

There is a tendency to regard applied behavioural research as less 'exciting' and less innovative than pure research, based on the notion that it consists solely of repetitive testing of ideas developed elsewhere. The recent progress in pest control and other applications shows that, on the contrary, applied studies involve an equal need to explore and extend original hypotheses about the organization of foraging decisions and the constraints on animals' feeding options. Also, the importance of repeatable effects in any practical management tool means that experimental testing of ideas in applied studies is often more rigorous, and hence a better test of success in understanding behaviour, than are some academic studies.

Many aspects of foraging theory are already used, explicitly or implicitly, in management. Some recent developments, however, have strong untapped potential. In particular, understanding the influence of stochastic variation in feeding conditions, and the ways in which animals adjust their behaviour to it (Stephens and Charnov, 1986), should help to improve many aspects of pest and wildlife management. Individual variation in foraging tactics is a further issue that has a major influence on the combined impact of feeding damage by pests, for example. Better appreciation of social transmission of information about food would also help to refine management of flocking species. Perhaps most urgently, we need to direct efforts to management of species in endangered habitats, to minimize the effect of future habitat loss.

REFERENCES

Arnold, G.W. and Dudzinski, M.L. (1978) *Ethology of Free-ranging Domestic Animals*, Elsevier, Amsterdam, 198.

Avery, M.L. (1984) Relative importance of taste and vision in reducing bird damage to crops with methiocarb, a chemical repellent. *Agriculture, Ecosystems and Environment*, **11**, 299–308.

Avery, M.L. (1985) Application of mimicry theory to bird damage control. *J. Wildlife Management*, **49**, 1116–21.

Bailey, P.T. and Smith, G. (1979) Methiocarb as a bird repellent on wine grapes. *Aust. J. Exp. Agric. Anim. Husbandry*, **19**, 247–50.

Barnard, C.J. and Thompson, D.B.A. (1985). *Gulls and Plovers. The Ecology and Behaviour of Mixed-species Feeding Groups*, Croom Helm, London.

Baudry, J. (1988) Hedgerows and hedgerow networks as wildlife habitat in agricultural landscapes, 111–24, in *Environmental Management in Agriculture* (ed. J.R. Park), Commission of the European Communities, Brussels.

Belovsky, G.E. (1984) Herbivore optimal foraging: a comparative test of three models. *Amer. Nat.*, **124**, 97–115.

Benjamini, L. (1980) Bait crops and mesurol sprays to reduce bird damage to sprouting sugar beets. *Phytoparasitica*, **8**, 151–61.

Blokpoel, H. (1976) *Bird Hazards to Aircraft*, Clarke, Irwin, Canada.

Boatman, N.D. and Sotherton, N.W. (1988) The agronomic consequences and costs of managing field margins for game and wildlife conservation. *Aspects of Applied Biology*, **17**, 47–56.

Brockmann, H.J. and Barnard, C.J. (1979) Kleptoparasitism in birds. *Anim. Behav.*, **27**, 497–514.

Brough, T. and Bridgman, C.J. (1980) An evaluation of long grass as a bird deterrent on British airfields. *J. Appl. Ecol.*, **17**, 243–53.

Brower, L.P. (1984) Chemical defences in butterflies. *Symp. Roy. Ent. Soc. Lond.*, **11**, 109–34.

Brown, C.R. (1986) Cliff swallow colonies as information centers. *Science*, **234**, 83–5.

Bryant, J.P. (1981) Phytochemical deterrence of snowshoe hare browsing by adventitious shoots of four Alaskan trees. *Science*, **213**, 889–90.

Buchsbaum, R., Valiela, I. and Swain, T. (1984) The role of phenolic compounds and other plant constituents in feeding by Canada geese in a coastal marsh. *Oecologia (Berl.)*, **63**, 343–9.

Bullard, R.W., Bruggers, R.L., Kilburn, S.R. and Fiedler, L.A. (1983) Sensory-cue enhancement of the bird repellency of methiocarb. *Crop Protection*, **2**, 387–98.

Bullard, R.W., Garrison, M.V., Kilburn, S.R. and York, J.O. (1980) Laboratory comparison of polyphenols and their repellent characteristics in bird-resistant sorghum grains. *J. Agric. Food Chem.*, **28**. 1006–11.

Bullard, R.W., York, J.O. and Kilburn, S.R. (1981) Polyphenolic changes in ripening bird-resistant sorghums. *J. Agric. Food Chem.*, **29**, 973–81.

Burn, A.J. (1988) Assessment of the impact of pesticides on invertebrate predation in cereal crops. *Aspects of Applied Biology*, **17**, 279–88.

Caraco, T. (1979) Time budgeting and group size: a test of theory. *Ecology*, **60**, 618–27.

Clark, W.R. (1987) Economics and marketing of Canada's 'capistrano', pp. 31–48, in *The Value of Birds* (eds A.W. Diamond and F.L. Filion), ICBP, Cambridge.

Conover, M.R. (1984) Responses of birds to different types of food repellents. *J. Appl. Ecol.*, **21**, 437–43.

Conover, M.R. (1985) Using conditioned food aversions to protect blueberries from birds: comparison of two carbamate repellents. *Appl. Anim. Behav. Sci.*, **13**, 383–6.

Cooper, S.M. and Owen-Smith, N. (1985) Condensed tannins deter feeding by browsing ruminants in a south African savanna. *Oecologia (Berl.)*, **67**, 142–6.

Dolbeer, R.A., Woronecki, P.P. and Stehn, R.A. (1982) Effect of husk and ear characteristics of resistance of maize to blackbird (*Agelaius phoeniceus*) damage in Ohio, USA. *Prot. Ecol.*, **4**, 127–39.

Draulans, D. (1987) The effectiveness of attempts to reduce predation by fish-eating birds: A review. *Biol. Conserv.*, **41**, 219–32.

Duncan, C.J. (1963) The response of the feral pigeon when offered the active ingredients of commercial repellents in solution. *Ann. Appl. Biol.*, **51**, 127–34.

Edmunds, M.E. (1974) *Defence in Animals: a Survey of Anti-Predator Defences*, Longman, Harlow, Essex.

Eiserer, L.A. (1980) Effects of grass length and mowing on foraging behavior of the American Robin (*Turdus migratorius*). *Auk*, **97**, 576–80.

Elgar, M.A. (1986) House sparrows establish foraging flocks by giving chirrup calls if the resources are divisible. *Anim. Behav.*, **34**, 169–74.

Elmahdi, E.M., Bullard, R.W. and Jackson, W.B. (1985) Calcium carbonate enhancement of methiocarb repellency for quelea. *Trop. Pest Manage.*, **31**, 67–72.

Eltringham, S.K. (1979) *The Ecology and Conservation of Large African Mammals*, Macmillan, London.

Feare, C.J. (1986) Pigeons: past, present and prerequisites for management. *Proc. 7th Brit. Pest Control Conference*, section 6, 1–14.

Feare, C.J., Greig-Smith, P.W. and Inglis, I.R. (1988) Current status and potential of non-lethal control for reducing bird damage in agriculture. *Proc. XIX Intern. Orn. Congr.*, I, 493–506.

Feare, C.J. and Inglis, I.R. (1979) The effects of reduction of feeding space on the behaviour of captive starlings *Sturnus vulgaris*. *Ornis Scand.*, **10**, 42–7.

Feare, C.J. and Wadsworth, J.T. (1981) Starling damage on farms using the complete diet system of feeding dairy cows. *Anim. Prod.*, **32**, 179–83.

Fox, G. and Linz, G. (1984) Evaluation of red-winged blackbird resistant sunflower germplasm. *Proc. Ninth Bird Control Seminar*, pp. 181–9, Bowling Green, Ohio.

Free, J.B. and Ferguson, A.W. (1983) Foraging behaviour of honeybees on oilseed rape. *Bee World*, **64**, 22–4.

Friedmann, H. (1955) The honey-guides. *US Natl. Museum Bull.*, **208**, 1–292.

Furness, R.W. and Monaghan, P. (1986) *Seabird Ecology*, Blackie, London.

Goss-Custard, J.D. (1980) Competition for food and interference among waders. *Ardea*, **68**, 31–52.

Gray, R. (1987) Faith and foraging: a critique of the paradigm argument from design, in *Foraging Behaviour* (eds A.C. Kamil, J.R. Krebs and H.R. Pulliam), Plenum Press, New York.

Greene, E. (1987) Individuals in an osprey colony discriminate between high and low quality information. *Nature*, **329**, 239–41.

Greig-Smith, P.W. (1987) Bud-feeding by bullfinches: methods for spreading damage evenly within orchards. *J. Applied Ecology*, **24**, 49–62.

Greig-Smith, P.W. (1988a) Hazards to wildlife from pesticide seed treatments, pp. 127–34, in *Application to Seeds and Soil* (ed. T.J. Martin), BCPC, Croydon.

Greig-Smith, P.W. (1988b) Wildlife hazards from the use, misuse and abuse of pesticides. *Aspects of Applied Biology*, **17**, 247–56.

Greig-Smith, P.W. and Wilson, G.M. (1984) Patterns of activity and habitat use by a population of bullfinches (*Pyrrhula pyrrhula*) in relation to bird-feeding in orchards. *J. Applied Ecology*, **21**, 401–22.

Guilford, T. and Dawkins, M.S. (1987) Search images not proven: a reappraisal of recent evidence. *Anim. Behav.*, **35**, 1838–45.

Halse, S.A. and Trevenen, H.J. (1985) Damage to medic pastures by skylarks in north-western Iraq. *J. Appl. Ecol.*, **22**, 337–46.

Henderson, M.T., Merriam, G. and Wegner, J. (1985) Patchy environments and species survival: chipmunks in an agricultural mosaic. *Biol. Conserv.*, **31**, 95–105.

Howe, H.F. (1984) Implications of seed dispersal by animals for tropical reserve management. *Biol. Conserv.*, **30**, 261–81.

Hudson, P.J. and Rands, M.R.W. (eds) (1987) *Ecology and Management of Gamebirds*, Blackwell Scientific Publications, Oxford.

Illius, A.W. (1986) Foraging behaviour and diet selection, pp. 227–37, in *Grazing Research at Northern Latitudes* (ed O. Gudmundsson), Plenum Press, New York.

Inglis, I.R. and Isaacson, A.J. (1978) The responses of dark-bellied brent geese to models of geese in various postures. *Anim. Behav.*, **26**, 953–8.

Krebs, J.R. (1974) Colonial nesting and social feeding as strategies for exploiting food resources in the great blue heron (*Ardea herodias*), *Behaviour*, **51**, 99–134.

Krebs, J.R., Stephens, D.W. and Sutherland, W.J. (1983) Perspectives in optimal foraging, pp. 165–216, in *Perspectives in Ornithology* (eds A.H. Brush and G.A. Clark), Cambridge University Press, Cambridge.

Lefebvre, L. (1983) Equilibrium distribution of feral pigeons at multiple food sources. *Behav. Ecol. Sociobiol.*, **12**, 11–17.

Lima, S.L., Valone, T.J. and Caraco, T. (1985) Foraging-efficiency-predation-risk trade-off in the grey squirrel. *Anim. Behav.*, **33**, 155–65.

Lima, S.L., Wiebe, K.L. and Dill, L.M. (1987) Protective cover and the use of space by finches: is closer better? *Oikos*, **50**, 225–30.

MacArthur, R.H. and Wilson, E.O. (1967) *The Theory of Island Biogeography*, Princeton Univ. Press, Princeton, NJ.

McKillop, I.G. and Sibly, R.M. (1988) Animal behaviour at electric fences and the implications for management. *Mammal Rev.*, **18**, 91–103.

McMillian, W.W., Wiseman, B.R. Burns, R.E. *et al.* (1972) Bird resistance in diverse germplasm of sorghum. *Agronomy Journal*, **64**, 821–2.

McNaughton, S.J. (1979) Grazing as an optimization process: grass-ungulate relationships in the Serengeti. *Amer. Nat.*, **113**, 691–703.

Mader, H.J. (1984) Animal habitat isolation by roads and agricultural fields. *Biol. Conserv.*, **29**, 81–96.

Margules, C., Higgs, A.J. and Rafe, R.W. (1982) Modern biogeographic theory: Are these any lessons for nature reserve design? *Biol. Conserv.*, **24**, 115–28.

Marshall, E.J.P. and Smith, B.D. (1987) Field margin flora and fauna; interaction with agriculture, pp. 23–33, in *Field Margins* (eds J.M. Way and P.W. Greig-Smith), Monograph 35, BCPC, Croydon.

Mason, J.R., Dolbeer, R.A., Arzt, A.H. *et al.* (1984) Taste preferences of male red-winged blackbirds among dried samples of ten corn hybrids. *J. Wildl. Manage.*, **48**, 611–16.

Mason, J.R. and Reidinger, R.F. Jr (1982) Observational learning of food aversions in red-winged blackbirds (*Agelaius phoeniceus*). *Auk*, **99**, 548–54.

Mason, J.R. and Reidinger, R.F. Jr (1983a) Generalization of and effects of pre-exposure on color-avoidance learning by red-winged blackbirds (*Agelaius phoeniceus*). *Auk*, **100**, 461–8.

Mason, J.R. and Reidinger, R.F. Jr (1983b) Conspecific individual recognition between starlings after toxicant-induced. sickness. *Anim. Learning Behav.*, **11**, 332–6.

Metcalfe, N.B., and Monaghan, P. (1987) Behavioural ecology: theory into practice, in *Advances in the Study of Behaviour*, **17**, 85–120.

Milinski, M. and Heller, R. (1978) Influence of a predator on the optimal foraging behaviour of sticklebacks (*Gasterosteus aculeatus* L.). *Nature*, **275**, 642–4.

Moss, R. (1972) Food selection by Red Grouse (*Lagopus lagopus scoticus* (Lath)) in relation to chemical composition. *J. Anim. Ecol.*, **41**, 411–28.

Murton, R.K., Coombs, C.F.B. and Thearle, R.J.P. (1972) Ecological studies of the feral pigeon *Columba livia* var. II. Flock behaviour and social organisation. *J. Appl. Ecol.*, **9**, 875–89.

Orians, G.H. and Pearson, N.E. (1979) On the theory of central place foraging, pp. 157–77, in *Analysis of Ecological Systems* (ed. D.F. Horn), Ohio State University, Columbus.

Owen-Smith, N. and Novellie, P. (1982) What should a clever ungulate eat? *Amer. Nat.*, **119**, 151–77.

Park, J.R. (ed.) (1988) *Environmental Management in Agriculture*, Commission of the European Communities, Brussels and Luxembourg. 250 pp.

Payne, C.C. (1988) Prospects for biological control, pp. 103–16, in *Britain since Silent Spring* (ed. D.J.L. Harding), Institute of Biology, London.

Pierce, G.J. and Ollason, J.G. (1987) Eight reasons why optimal foraging theory is a complete waste of time. *Oikos*, **49**, 111–18.

Potts, G.R. (1986) *The Partridge: Pesticides, Predation and Conservation*, Collins, London.

Potts, G.R. and Vickerman, G.P. (1974) Studies on the cereal ecosystem. *Adv. Ecol. Res.*, **8**, 107–97.

Rands, M.R.W. (1985) Pesticide use on cereals and the survival of grey partridge chicks: a field experiment. *J. Appl. Ecol.*, **22**, 49–54.

Rands, M.R.W. (1986) The survival of gamebird (*Galliformes*) chicks in relation to pesticide use on cereals. *Ibis*, **128**, 57–64.

Rasa, O.A.E. (1984) Dwarf mongoose and hornbill mutualism in the Taru Desert, Kenya. *Behav. Ecol. Sociobiol.*, **12**, 181–90.

Reidinger, R.F. Jr and Mason, J.R. (1983) Exploitable characteristics of neophobia and food aversions for improvements in rodent and bird control, pp. 20–39, in *Vertebrate Pest Control and Management Materials, Fourth Symposium* (ed. D.E. Kaukeinen), ASTM, Philadelphia.

Rhoades, D.F. and Cates, R.G. (1976) Towards a theory of plant anti-herbivore chemistry. *Rec. Adv. Phytochem.*, **10**, 168–213.

Robbins, C.T. (1983) *Wildlife Feeding and Nutrition*, Academic Press, London, 343.

Robbins, R.J. (1981) Considerations in the design of test methods for measuring bait shyness, pp. 113–23, in *Vertebrate Pest Control and Management Materials: Third Conference* (eds E.W. Schafer Jr and C.R. Walker), ASTM, Philadelphia.

Rogers, J.G. Jr (1978) Some characteristics of conditioned aversion in red-winged blackbirds. *Auk*, **95**, 362–9.

Rooke, I.J. (1984) Methiocarb-induced aversion to grapes by grey-backed white-eyes. *J. Wildl. Manage.*, **48**, 444–9.

Salvi, A. (1987) Utilisation de perchoirs dans la recherche alimentaire par des bandes d'etourneaux sansonnets (*Sturnus vulgaris*). *Biol. Behav.*, **12**, 12–27.

Savory, C.J. (1978) Food consumption of red grouse in relation to the age and productivity of heather. *J. Anim. Ecol.*, **47**, 269–82.

Schafer, E.W., Bowles, W.A. and Hurlbut, J. (1983) The acute toxicity, repellency and hazard potential of 998 chemicals to one or more species of wild and domestic birds. *Arch. Environ. Contam. Toxicol.*, **12**, 355–82.

Seeley, T.D. (1985) *Honeybee Ecology*, Princeton Univ. Press, Princeton, NJ.

Smith, N.G. (1969) Provoked release of mobbing — a hunting technique of *Micrastur* falcons. *Ibis*, **111**, 241–3.

Sotherton, N.W. (1985) The distribution and abundance of predatory Coleoptera overwintering in field boundaries. *Annals of Applied Biology*, **106**, 17–21.

Stephens, D.W. and Krebs, J.R. (1986) *Foraging Theory*, Princeton Univ. Press, Princeton, NJ.

Stephens, D.W. and Charnov, E.L. (1986) Optimal foraging: some simple stochastic models. *Behav. Ecol. Sociobiol.*, **10**, 251–63.

Stevens, J. and De Bont, A.F. (1980) Choice by starlings (*Sturnus v. vulgaris* L.) among different cherry cultivars. *Agricultura (Heverlee)*, **28**, 421–36.

Stewart, C.N., Reidinger, R.F. Jr and Mason, J.R. (1983) A method for inferring the taste qualities of rodenticides to rodents, pp. 155–64, in *Vertebrate Pest Control and Management Materials, Fourth Symposium* (ed. D.E. Kaukeinen), ASTM, Philadelphia.

Summers, D.D.B. and Huson, L.W. (1984) Prediction of vulnerability of pear cultivars to bullfinch damage. *Crop Protection*, **3**, 335–41.

Sullivan, T.P., Nordstrom, L.O. and Sullivan, D.S. (1985) Use of predator odors as repellents to reduce feeding damage by herbivores. I. Snowshoe hares (*Lepus americanus*). *J. Chem Ecol.*, **11**, 903–19.

Sullivan, T.P. and Sullivan, D.S. (1982) The use of alternative foods to reduce lodgepole pine seed predation by small mammals. *J. Applied Ecology*, **19**, 33–45.

Sunderland, K.D. (1975) The diet of some predatory arthropods in cereal crops. *J. Appl. Ecol.*, **12**, 507–75.

Sutherland, W.J. and Parker, G.A. (1985) Distribution of unequal competitors, pp. 255–73, in *Behavioural Ecology: Ecological Consequences of Adaptive Behaviour* (eds R.M. Sibly and R.H. Smith), Blackwell Scientific Publications, Oxford.

Tahvanainen, J., Julkunen-Tiitto, R. and Kettunen, J. (1985) Phenolic glycosides govern the food selection pattern of willow feeding leaf beetles. *Oecologia (Berl.)*, **67**, 52–6.

Temple, S.A. (ed.) (1978) *Endangered Birds: Management Techniques for Preserving Threatened Species.* Univ. of Wisconsin Press, Madison.

Thompson, R.D., Grant, C.V. and Elias, D.J. (1981) Factors affecting red-winged blackbird response to methiocarb, an avian repellent. *Pesticide Biochem. Physiol.*, **15**, 166–71.

Valiela, I. and Rietsma, C.S. (1984) Nitrogen, phenolic acids, and other feeding cues for salt marsh detritivores. *Oecologia (Berl).*, **63**, 350–6.

Way, J.M. and Greig-Smith, P.W. (eds) (1987) *Field Margins*, Monograph No 35, BCPC, Croydon.

Werner, E.E., Gilliam, J.F., Hall, D.J. and Mittelbach, G.G. (1983) An experimental test of the effects of predation risk on habitat use in fish. *Ecology*, **64**, 1540–8.

Woolhouse, M.E.J. (1983) The theory and practice of the species-area effect, applied to the breeding birds of the British woods. *Biol. Conserv*, **27**, 315–32.

Wright, E.N. (1980) Chemical repellents — a review, pp. 164–72, in *Bird Problems in Agriculture* (eds E.N. Wright, I.R. Inglis and C.J. Feare), BCPC, Croydon.

3

Social behaviour

PAT MONAGHAN

The social environment in which animals live varies greatly between species and, in some cases, the same species in different environments will adopt a different social organization. Like any other feature of an animal's morphology or behaviour, social organizations will be shaped by the forces of natural selection to maximize the individual's genetic contribution to the next generation. This will mainly come about through maximizing the number of offspring an animal produces in its lifetime; however, where the opportunities for an individual to reproduce are constrained, perhaps due to a shortage of mates or breeding sites, assisting in the reproductive attempts of close relatives may be favoured by the process of kin selection (see Krebs and Davies, 1984 for a full discussion). All interactions between animals of the same species come under the umbrella of social behaviour, which therefore includes the mating patterns, group life, dispersal and dominance relationships between conspecifics. When trying to manage the behaviour of animals, whether to control population sizes, maximize productivity or improve welfare, we must take into account their social behaviour, and the social environment to which this behaviour is adapted. In this chapter we shall examine some of the ways in which an understanding of social behaviour can overcome problems in improving productivity of species kept in captivity for commercial and conservation purposes, and in controlling animal populations in the wild whether they be pests or endangered species.

3.1 SPACING PATTERNS

Animals very rarely space themselves randomly with respect to other individuals of the same species, to which they may be either attracted or repelled. Spacing patterns vary from the defence of exclusive territories, as seen in young salmonids in streams (see Chapter 5), to large, cohesive flocks formed by many birds in winter. This variation can be interpreted as a consequence of variation in both the distribution and predictability of key resources. Territoriality is only favoured when the benefits of exclusive use of a resource outweigh the costs of defending it; that is, the resource is 'economically defendable' (see Davies and Houston, 1984 for a full discussion of the costs and benefits of territoriality). Thus, when food is distributed in large, unpredictable patches, we should not expect to find animals defending feeding territories. On the other hand, where food is more scattered and predictable, it may be worth defending a territory in order to secure access to the food and exclude competitors. For example, many woodland birds which feed on insects during the breeding season defend territories, whereas seabirds, which feed on shoaling fish, do not defend feeding territories but rather tend to feed in flocks.

This approach based on the consideration of costs and benefits can be used to predict whether or not animals should live in groups. The advantages of group living have been well reviewed by Bertram (1978) and include improved defence against predators, reduced vigilance costs leaving more time available for feeding, improved location of food and reduced chance of being the victim should a predator strike. Costs largely arise from increased competition, and the effect of group feeding on food intake will thus depend on the nature and abundance of the food and the foraging method employed. Figure 3.1, which shows the intake rates of starlings (*Sturnus vulgaris*) feeding at a cattle trough in groups of different sizes, clearly demonstrates these effects. Initially, as the group size increases, the intake rate of the birds increases since, as there are more individuals to look out for predators, more time can be devoted to feeding. However, once the group size increases above about seven birds, competition for access to the restricted feeding space at the trough causes a reduction in intake rate (Feare, 1984).

3.1.1 Controlling the level of competition

Individual spacing patterns can cause considerable problems when animals are reared commercially in either captive or wild conditions (see Chapter 7). The aim in such rearing programmes is usually to get the maximum density and/or growth rates at a minimum cost to the producer. This is the case whether one is managing a pig unit, a grouse moor or a fish farm. It is not simply a case of putting together the maximum number of animals with the estimated amount of food required. The social behaviour of the animals will mean that dominants may exclude subordinates from feeding areas, and in some cases the mere sight of a dominant will suppress feeding, and even

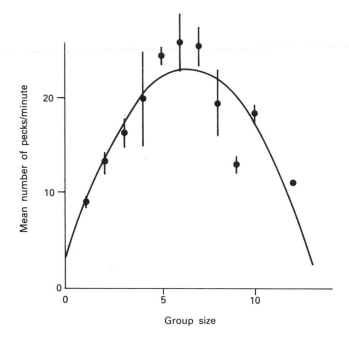

Figure 3.1 The feeding rate of starlings at a cattle trough in groups of different sizes. (After Feare, 1984).

maturation and reproduction, in subordinates (see below and Chapters 5 and 6).

Territorial behaviour may prevent some animals from being established in an area at all. For example, juvenile stream-living salmonids are often territorial, and experimental manipulations have demonstrated that the size of territories is dependent on food supply, not on population size. Adding more fish to a stream will not reduce the average territory size. Slaney and Northcote (1974) showed that adding more juvenile rainbow trout (*Salmo gairdneri*) to a stream simply resulted in subordinate fish being forced to migrate downstream (Figure 3.2a). Decreasing food abundance caused dominant fish to increase their territory size, thereby producing an increased migration of subordinates (Figure 3.2b). These effects occurred within hours of fish being added. This is of obvious importance when it comes to restocking streams; the numbers of residents cannot be increased beyond the level determined by territoriality, and to change this the food supply must also be manipulated. However, one benefit of this spacing behaviour of the fish is that the growth rates of those

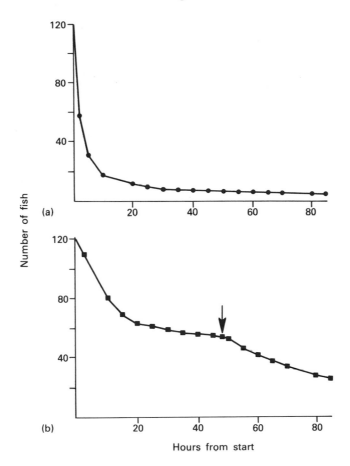

Figure 3.2 The numbers of juvenile trout remaining in sections of two experimental streams after initial stocking. In (a) no food was added; in (b) food was added until the point marked by the arrow after which it was withdrawn. (After Slaney and Northcote, 1974.)

fish which obtain territories will not be reduced if the stream is overstocked. Without territorial behaviour, overstocking might lead to reduced growth due to increased competition. This has in fact been found to occur in salmonid fish in lakes; such fish do not defend territories, since their food supply is very patchily distributed. Overstocking in lakes produces a stunted population whose growth rates are poor (Johnston, 1965).

A similar kind of problem occurs in the management of gamebirds. The extent to which the bird population will respond to management practices is limited by the social behaviour of the birds. Red grouse (*Lagopus lagopus scotius*), one of the most intensively studied game-birds, are highly territorial and monogamous. In autumn, males compete for territories on heather moorland and successful males then pair with females. Both males and females which do not obtain a territory remain in groups, and these birds have usually died or emigrated by the following spring. Variations in food, cover, weather, predation, parasitism and disease have all been shown to affect grouse density, largely through effects on the territorial behaviour of the males (Moss and Watson, 1984; Hudson and Rands, 1988). The territory is used for feeding and usually, but not always, to rear the young. Red grouse selectively eat heather shoots rich in nitrogen and phosphorus. Manipulating the quality of the food will influence territory size, and this is the main tool the gamekeeper has to influence the number of birds breeding in the area. However, the quality of the food must be manipulated over the entire area occupied by the population being managed, since the territorial behaviour of the birds means that the provision of high quality patches would only affect those pairs within whose territories the patches lay. This contrasts with other non-territorial gamebirds such as pheasants (*Phasianus colchicus*), for which central feeding stations can be set up to increase numbers in a particular area (Avery and Ridley, 1988), though of course differences in the competitive abilities of birds may influence how much of the supplementary feed they obtain.

It is not surprising that animals differ in their competitive abilities, yet many commercial rearing systems seem to be based, at least initially, on the assumption that resources will be shared equally between all individuals (see Chapters 5 and 6). Domestic animals are thus often forced to feed in conditions where competition is intense. For example, the common agricultural practice of locally supplementing the food of livestock rarely takes into consideration the effects this may have on competition and the use of space by the animals involved. When feed blocks are used to supplement the forage available to sheep in grazing areas, comparatively large blocks are usually left in a very small number of patches. Competition at the blocks results in levels of aggression which are abnormally high for a grazing animal, a situation which could cause chronic stress in subordinates. In addition, the animals may reduce the extent of their ranging to the neighbourhood of the blocks. This leads to overgrazing and trampling of these areas and undergrazing of the rest, thus exacerbating the original problem of shortage of forage (Lawrence and Wood-Gush, 1988). One solution to the problem with feed blocks would be to spread the resource out more thinly by decreasing the size of the blocks and increasing the number of depots. This would reduce the economic defendability of the resource and so reduce the level of competition.

The design of standard animal feeding troughs gives rise to similar problems. Domestic pigs kept in groups soon form hierarchies (see Chapter 6), and the common procedure of feeding them from open troughs presents the dominant pigs with a defendable food supply which they are able to monopolize. In more natural conditions, such resource defence rarely occurs, since pigs are usually feeding on dispersed food such as grass and roots and feed several metres apart. Stolba (1985) carried out a series of behavioural experiments in which he varied the way in which the food was presented to a group of pigs. By monitoring the behaviour of different individuals he was able to pin-point the conditions which minimized differences in feeding rates between dominants and subordinates. Increasing the length of the troughs decreased their defendability and allowed subordinates to feed without being in close contact with dominants; however, dominant pigs were still able to deter other pigs from feeding by facial threat displays. Stolba then tried various ways of screening off the feeding pigs from each other (Figure 3.3). The most cost-effective

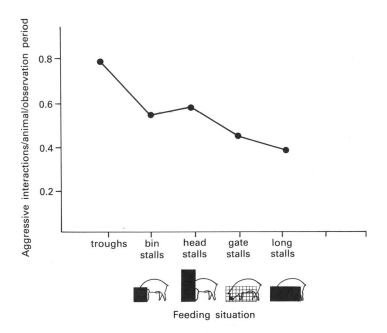

Figure 3.3 Frequency of aggressive interactions per pig during a 20 minute observation period of 10 pigs in different feeding situations.

method, in terms of financial outlay and behavioural results, was to divide the trough into individual sections using small screens placed along its length (bin stalls). These screens effectively prevented each pig at the trough from seeing the head of its neighbour, and reduced both the incidence and effect of threat posturing by dominants (Figure 3.3). A simple and practical solution to a complex problem was thus provided by careful study of the behavioural causes.

Differences in the feeding localities of dominant and subordinate animals may also influence their capacities to carry disease, as is the case with the herring gull (*Larus argentatus*). In feeding interactions, female herring gulls tend to be subordinate to their larger male counterparts and, as a consequence, are more likely to occupy the less preferred feeding areas (Greig *et al.*, 1985). In Scotland, while female herring gulls outnumber males at rubbish tips, they are often concentrated on the older, more putrid refuse. As a result, more than double the proportion of females compared with males carry the Salmonellae, bacteria which cause food poisoning in man and domestic animals (Table 3.1). These birds may be involved in the transport of these pathogens between different areas (Monaghan *et al.*, 1985).

Table 3.1 The proportion of adult herring gulls of both sexes carrying Salmonellae in winter

	Males	*Females*
No. carrying Salmonellae	10	18
No. not carrying Salmonellae	111	73
% carrying Salmonellae	8.0	20.0

(After Monaghan *et al.*, 1985)

3.1.2 Controlling group structure

Many animals live in complex social groups, and thrive only if the normal group structure is maintained. This must be taken into account whether one is keeping bees, conserving animals in the wild or maintaining groups in zoos. Keeping social animals in isolation in captivity can give rise to many behavioural problems (see Chapters 6 and 7). Physical problems can also arise where, for example, there is no opportunity for social grooming; increased incidence of ectoparasites may result, as has been demonstrated in mice which can develop skin problems in the head and neck region when kept alone (Frazer and Waddell, 1974). Young animals may learn important skills from contact with older individuals and early social experience can have important effects on subsequent social and sexual preferences through the mechanisms of filial and sexual imprinting (see Bateson, 1978a and Ten Cate, 1989 for reviews and also Chapter 8). However, the solution is not simply a matter of not keeping animals alone which normally live in groups. The age and sex composition of the group must also be considered. When housing animals in captivity, great care must be taken over the group structure, since the consequences of social stress can be very severe. For example, in wild talapoin monkeys (*Miopithecus talapoin*), males and females usually live in separate groups; males may be killed by females if the two sexes are housed together in captivity (Kleiman, 1980). The effects of social stress are particularly marked in tree shrews (*Tapaia belangeri*) which have been studied in detail (von Holst, 1974). These animals are highly territorial in the wild, and normally live either singly or in mated pairs with their small young. In captivity, where there is no opportunity for sexually mature young to disperse, mature individuals of the same sex fight and dominance relationships are quickly established. If continually housed in visual contact with a dominant the subordinate individual shows signs of severe stress and generally die as a result of renal failure. However, if separated from the dominant early enough, subordinate individuals recover. Therefore, if tree shrews are kept in captivity, which they often are, extreme care must be taken to prevent such stress symptoms developing by removing young animals from the family groups before they reach sexual maturity.

In animals bred in captivity, the timing of separation of mother and young is important to both productivity (Chapter 6) and welfare (Chapters 7 and 8). Many young primates, for example, have a long period of social dependency which is important to their social development. In chimpanzees, *Pan troglodytes*, weaning does not occur until around the 5th year (Clark, 1977) and the average birth interval is almost six years. In captivity, premature separation of mothers and infants can have profound effects; juvenile chimps as old as eight years may grieve to death as a result of being orphaned (McGrew, 1981). Subsequent reproductive behaviour may be adversely affected by early separation from the mother, as has been shown in Japanese monkeys (*Macaca fuscata*) (Hasegawa and Hiraiwa, 1980). Most primate social life is based on kinship, and in many species several generations of related individuals form the core of the group's structure. Such relationships are favoured by kin selection, and the more

closely related the individuals, the more co-operation between them
(Kurland, 1977). The ability to interact with kin seems to be essential
to the normal social development of many primate species (McGrew,
1981).

Such considerations do not apply only to species kept in captivity.
In conserving rare or threatened species in the wild, it is important
that management policies are geared towards encouraging the right
social environment necessary to maintain a healthy population. The
chough (*Pyrrhocorax pyrrhocorax*) is a comparatively rare bird in
Britain. Formerly known as the Cornish chough, it is now sadly extinct
in England. There is however a good deal of interest in re-establishing
the bird in Cornwall. However, in addition to ensuring that the necessary habitat is available to support the birds, it is important that the
required social organization is promoted if a self-sustaining population
is to be established after the initial release of birds. Like many corvids,
choughs are highly social birds and young birds in particular feed and
roost communally (Still *et al.*, 1986). The anti-predator and foraging
advantages of such group living may be essential to the survival of the
young birds, and may also provide the opportunity to establish pair
bonds; in addition, as in many other species, young birds may learn
many of the skills essential for survival by associating with older individuals, and a lack of experienced birds in a population could result
in poor survival (Monaghan, 1989). Thus in designing any re-establish-
ment programme due consideration must be given to the age composition and size of the initial group, the likely production of young
and opportunities for flock formation.

3.1.3 Controlling dispersal

In many species, young animals disperse from their area of birth and
breed in new localities. In mammals, such dispersal is predominantly
by males while in birds it is female biased, though there are exceptions in both cases (Greenwood, 1980). Animals disperse in order to
maximize their chances of survival and reproduction. They may be
moving to avoid local competition, to avoid inbreeding or to gain
information of the availability of potential mates, feeding grounds or
breeding sites (Baker, 1978). Differences in the competitive abilities
and dominance status of animals, in altering their spacing and
dispersal patterns, may also have consequences for their potential to
become pests. Young animals are often less able to secure territories
and may wander more widely over the available habitat. A high
proportion of the 'nuisance' bears in American national parks are
males which have not managed to secure a territory; these bears

range widely in search of food and come into more contact with humans as a result (McArthur, 1981). In foxes, (*Vulpes vulpes*), it is predominantly the males which disperse; young males disperse more widely than adults, particularly small cubs from large litters which are presumably subordinate (Harris and Trewhella, 1988). This dispersal is seasonal, and understanding how, when and why foxes move is very important in developing a strategy to contain a rabies outbreak should one occur (Trewhella *et al.*, 1988).

In primate groups, it is generally the case that some members leave their natal group and join other groups. As in other animals, this probably serves to reduce inbreeding depression in the wild (see below). In most primate species living in multi-male groups, it is the adolescent males that leave at puberty, though in some monogamous species like marmosets, offspring of both sexes may leave (Packer, 1975; McGrew, 1981). Many of the problems which arise in species kept in zoos result from a lack of opportunity for appropriate dispersal. In captive lion tamarins, (*Leontopithecus rosalis*), for example, sub-adult males may be killed by their mother as they approach sexual maturity if they continue to be housed together (Kleiman, 1980). In the wild, these young males would normally disperse at this time. In chimpanzees on the other hand, it is the young females which transfer between groups (Pusey, 1980). Exchanges between zoos must therefore be based on movement of the appropriate age and sex class of animals in order to minimize disruption of the groups and ensure acceptance of the new individuals. Failure of rhesus monkeys (*Macaca mulatta*) to breed in a captive colony over several years was later shown to have arisen because young females were moved between groups, rather than the young males which would normally disperse (McGrew, 1981).

3.1.4 Controlling predation risk

While the optimum solution for a sheep farmer may be to have sheep dispersed in order to even out grazing pressure, the opposite may be true for anyone rearing animals which suffer a high predation risk. Numerous studies have shown that, in animals which do not rely on concealment as a protection against predators, the risk of predation tends to be much higher for solitary individuals than for members of a group (Bertram, 1978; Pulliam and Caraco, 1984). For example, Kenward (1978) found that a much greater proportion of attacks by a goshawk (*Accipiter gentilis*) were successful when directed against solitary wood pigeons (*Columba palumbus*) compared with pigeon flocks (Figure 3.4). This was largely due to earlier detection of the

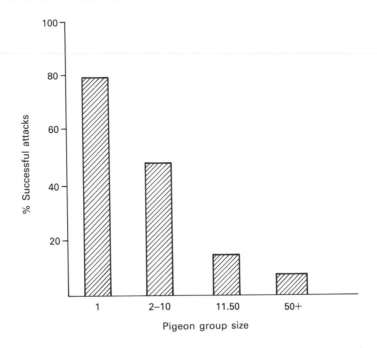

Figure 3.4 The proportion of attacks by a goshawk which were successful when directed at pigeons in different group sizes.

predator by the pigeon groups. Neill and Cullen (1974) found that a variety of predators were much less successful when faced with fish schools than when attacking solitary fish. This time the effect seems to be largely due to confusion of the predator which, in the presence of a throng of rapidly moving potential victims, is unable to concentrate on any one for long enough to strike at it successfully.

Thus if some members of a population are more solitary than others, they may be more vulnerable to predation. This effect is seen in the vulnerability of gamebirds to human predators. Terrestrial male red grouse are relatively solitary and are more vulnerable in driven shoots, since the shooters appear to be able to take better aim (Robertson and Rosenberg, 1988). This means that shooting may take a heavier toll of high quality territorial individuals which are replaced by lower ranking individuals from flocks, to the detriment of the breeding stock (Moss and Watson, 1984). In grey partridges (*Perdix perdix*) males are similarly more solitary and therefore more vulnerable to shooting (Robertson and Rosenberg, 1988). However, since grey partridges have a polygamous mating system, the effect on the breeding population may be less, provided all of the top ranking males are not lost.

The tendency of animals to group is frequently exploited by man as a predator; while schooling in fish is an effective strategy against natural predators, it offers no protection against fishing nets, which, if skilfully managed, catch the entire shoal. For fishermen then, schooling makes fishing more cost-effective, and commercially important shoaling species, such as the herring (*Clupea harangus*), present extremely profitable patches to the human predator, provided the shoals can be located. The development of efficient school-finding fishing technology such as sonars makes shoal location increasingly easy, which means that search time tends to be short once the fishing grounds are reached. This has the unfortunate effect that, even when a fish population is in serious decline, entire shoals will continue to be caught. This is in contrast to more natural predation where, once the prey becomes less abundant, the predator finds it more and more difficult to locate and may switch to alternative prey. Such density dependent effects means that predators rarely drive their prey populations to extinction. However, this is not the case with human predation. The implication for fisheries management is that there may be very little warning of any decline in fish stocks until the population has reached very low levels (Murphy, 1980). A depressing example of this is the Atlanto-Scandian herring fishery, which crashed from a catch of 1.7 million tons in 1966 to just 21000 tons within four years, due to the virtual elimination of the fish.

Where the aim is to reduce predation by altering the behaviour or abundance of the predator, a knowledge of the social behaviour of the predator is equally important. For example, in North America, coyotes (*Canis latrans*) are important predators of livestock (see Chapter 6). Both male and female coyotes defend overlapping territories, and sometimes a territory is occupied by a group. Studies of the females in a population using radio tracking showed that around 66% were territorial, occupying areas of 2–3km^2, and the remaining 34% were transients, wandering over larger areas of over 12km^2 (Windberg and Knowlton, 1988). Males similarly can be either residents or transients (Gese *et al.*, 1988). Territory size varies in relation to habitat, particularly the amount of cover, and the ranges of transients usually overlap with the ranges of residents and other transients (Gese *et al.*, 1988). The territorial residents are largely adults while the transients are generally young animals 1–2 years old. An understanding of these spacing patterns has important implications for coyote management policies; if control is to be attempted, coyotes need to be removed over a wide area since there is a reservoir of highly mobile transients available to take up any vacated territories. Coyotes mainly kill live-stock when they are feeding young. A more effective management policy may therefore be to sterilize territory holders within a problem area. These animals would then continue to defend the area against incoming coyotes, but the predation on livestock would be reduced since they would produce no young (Till and Knowlton, 1983).

3.2   CONTROLLING THE SIZE OF THE BREEDING POPULATION

There is a frequent need to manage the numbers of animals, for example in controlling pests, conserving rare species or managing game numbers. An understanding of the social organization and life history strategies of the animals involved is essential if management policies are to be effective in controlling the population size. There is, for example, little point in culling female foxes to reduce the popula-tion size in an attempt to protect livestock. Dominant female foxes can suppress reproduction in other females (MacDonald, 1980). Culling breeding females thus simply serves to remove this constraint on the reproductive output of subordinates, and production of young may not be reduced unless very persistent culling takes place.

Social behaviour can have very considerable influences on popu-lation density. For example red grouse numbers in an area are largely controlled by the territorial behaviour of the males. While the function of the territorial behaviour is to maximize individual reproductive success, a consequence of the competition between individuals is the limitation of breeding numbers in the area. It does not necessarily follow that shooting grouse will reduce numbers breeding, since the birds which fail to secure territories can act as a reservoir to fill vacant areas, in a manner similar to the coyotes already mentioned. However, the population structure may be altered. In beaver populations un-exploited by man, young beavers usually disperse before being able to breed, since there are few vacant territories available. Since dispersal is risky however, this favours delayed maturation and the young beavers are experienced foragers by the time they leave the natal area. The average age of first breeding for females is four years. In contrast, in a beaver population exploited by fur trappers, the reduced life expectancy of adults means that territories become vacant more frequently. In such a population there is much less dispersal of young animals. In addition, the availability of local breeding sites favours early maturation, with females beginning to breed a year earlier than in unexploited populations (Boyce, 1981). While the two types of population have different age structures, the density of beavers is not

affected. These data are important in assessing the level of exploitation which beaver populations can withstand.

In general, young animals suffer higher mortality rates than adults.
An increased mortality in breeding adults is likely to have more
dramatic effects on population size than the loss of younger age
classes; once the level of adult mortality exceeds the level of recruitment of young the population will begin to fall. The number of
breeding females is the important limiting factor, while the importance
of males in the population will vary with the mating system. Culling of
breeding adults of both sexes was the management strategy introduced in an attempt to reduce the numbers of the monogamous
herring gull breeding in many areas of Britain. This bird increased
dramatically during much of the present century up to the 1970s. On
the Isle of May in the Firth of Forth, culling was first started in 1972 to
protect the island's vegetation. Over ten years the breeding population
was reduced from around 14000 pairs to 3000, but over 40000 birds
had to be culled to achieve this end. This management policy has
removed several of the constraints on the reproductive rate of the
remaining birds and changed the social structure of the colony. The
preferred breeding sites of young herring gulls recruiting into a colony
are in the densest breeding areas; fierce competition for these sites
occurs since the low annual adult mortality means that normally few
such sites become available each year (Chabrzyk and Coulson, 1976).
In an undisturbed large colony, young birds unable to obtain a central
site may either defer breeding or move to the periphery where
breeding density is low and predation risk high. When density is low,
recruitment is low since few young birds are attracted. When density is
high, recruitment is again low since, although young birds try to obtain
sites in these areas, they are repelled by the established breeders. The
level of recruitment will therefore only be that of the adult mortality
rate. Recruitment is highest at the intermediate densities. If a high
density colony is reduced to intermediate density by culling, this
optimizes the conditions for recruitment. As a result, large numbers of
birds will have to be culled each year, since vacant territories are
constantly being filled. This appears to be what happened on the Isle
of May, since the previously very dense colony was thinned out
overall by culling. The increased opportunities for young birds to
obtain breeding territories resulted in a drop from 5.6 to 4.3 years in
the mean age of first breeding (Coulson *et al.*, 1982). A more effective
management strategy would have been to clear parts of the island
completely of nesting gulls. Few young birds would then attempt to
breed in these areas, and their re-colonization could be prevented by
nest clearance. The remaining areas would continue to have a high

density of breeding birds, but recruitment would be fairly low and any peripheral spread could again be prevented by clearance. An additional effect of the reduction in numbers of herring gulls breeding on the Isle of May appears to have been a reduction in competition, since the young birds returning to breed appear to have attained a larger size and birds breeding there now lay larger eggs (Coulson *et al.*, 1982). Since there is considerable movement of young herring gulls away from their colony of birth to breed, the reduced young production resulting from culling at a large colony can have marked effects on the level of recruitment into colonies elsewhere. Furthermore, the overall pattern of recruitment may be disrupted by the disturbance caused at the culled colony (Coulson, in press).

Under a monogamous mating system like that of the herring gull, males and females are equally important and (provided the sex ratio is equal) a loss of either sex will reduce the productivity of the population. Where polygyny occurs however, the majority of matings tend to involve only a small number of males, and thus many males do not breed (Figure 3.5). The sex ratio can be skewed in favour of females without reducing the reproductive output of the population since all females can still be mated. In harvesting populations of the polygynous pheasant, one may therefore wish selectively to take out adult males, many of which are, so to speak, surplus to the females' requirements! Population size can be maintained despite shooting pressure. Such a selective shooting policy would, however, be applicable to the monogamous red grouse (Avery and Ridley, 1988).

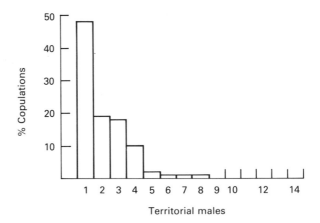

Figure 3.5 The percentage of the total number of observed copulations obtained by different male grouse (*Lyrurus tetrix*) at a lek. The males have been ranked in order of mating success. (After Kruijt and Hogan, 1967.)

3.3   CONTROLLING MATING BEHAVIOUR

Manipulation of mating behaviour is desirable in a wide variety of contexts, including improving the productivity of commercial livestock (Chapter 6), captive breeding of endangered species (Chapter 7) and control of pests by reducing their reproductive rate. The development of behavioural ecology has increased our understanding of the mating strategies of animals, which can largely be viewed in terms of the relative investments of males and females in the reproductive attempt, with each sex trying to get the maximum reproductive success for the minimum investment. That females produce large and comparatively costly eggs, while males produce small and relatively inexpensive sperm, means that there is an initial asymmetry in investment. The reproductive output of females is limited by the resources available to produce eggs, whereas that of males is limited by the number of females with whom they can successfully mate. We should thus expect females to go for quality and males for quantity when choosing mates! Of course this basic pattern is modified by the extent to which one parent can successfully rear the young. Males only benefit from taking another mate if their reproductive output overall is increased. Theoretical considerations thus suggest that, unless the time and resources required for breeding are minimal, it is advantageous for individuals to discriminate between potential mates on the basis of factors which will affect their reproductive success (Partridge and Halliday, 1984). Mate choice is thus likely to be an important component of the reproductive behaviour of many species.

Captive breeding is an important conservation tool. A recent survey suggests that around 2000 species of terrestrial vertebrates, including 160 primates, 100 large carnivores and 800 birds, may have to be bred captively in zoos to save them from extinction (Read and Harvey, 1986). Yet many captive breeding programmes begin with the assumption that simply putting a sexually mature male and female together in the same cage at the right time will give rise to a family. The saga of the failed matings of captive giant pandas has emphasized that this is far from the case. An important component of the mating behaviour of many species, and one that is often denied them in captivity, is courtship. The most elaborate and spectacular behaviour patterns produced by animals are part of the courtship displays performed by males to attract mates, and the fiercest fighting that occurs is between rival males contesting over females. These two aspects of sexual selection, which Darwin termed the power to charm the females and the power to conquer other males in battle, have given rise to behaviours as diverse as the strutting of peacocks and the rutting of stags. Courtship displays were traditionally thought by ethologists to function largely in bringing the male and female into breeding condition, and were seen as a harmonious ritual which preceded actual mating. However, studies in behavioural ecology have now shown that courtship is not quite so harmonious as it may appear; an important function of courtship displays is mate assessment. For example, in common terns (*Sterna hirundo*), males court females by bringing fish, which they hold in their bills and display to potential mates. In the later stages of courtship, once the pair bond is formed, these fish are eaten by females and contribute to the reserves they require to produce the clutch of eggs. In the early stages however, the display fish carried by a male seem to be used by females as a cue to how good he will be at foraging for fish when chicks are hatched. Nisbet (1973) found that the foraging performance of males during courtship was a good indication of their foraging performance at the chick phase. It thus pays females to mate with males carrying high quality display fish, and females responding in this way will have been favoured by natural selection. Without the opportunity for such mate assessment, some animals will not breed at all or have poor success. A number of studies have shown that animals given an opportunity to choose their mates have increased reproductive rates (Partridge, 1980). Thus an animal husbandry system that arbitrarily creates pairs may produce sub-optimal results.

A study of canvasbacked ducks (*Aythya valisineria*) by Bluhm (1985) clearly illustrates the importance of courtship and mate choice. These ducks, like many waterfowl, normally pair in their wintering areas after a prolonged period of courtship. Once a pair is formed, they then remain close together, with the male defending the female's feeding territory and the female chasing off other males. When paired ducks were separated from their mates for over a week and then put together in a large flock, Bluhm found that the same pairs re-formed. In some cases this occurred within only fifteen minutes, indicating good individual recognition. When canvasbacks were taken from a flock of displaying birds and randomly assigned to pairs in breeding cubicles, birds that had previously been displaying stopped when forced into pairs in this way. None of these birds produced eggs (Table 3.2). Those females which had already been paired at the time the forced pairing was introduced behaved so aggressively to their new mates that many of these males actually died. In contrast, already formed canvasback pairs which were put into the cubicles together continued to display and went on to produce eggs (Table 3.2). Thus both the break-up of existing pairs and the lack of mate choice can give rise to breeding failure, and the stress produced by forced

Table 3.2 The breeding success of canvasback ducks in relation to pair status

Pair status	% females aggressive	% females laying eggs	n
Natural pairs	5	89	19
Forced type I	17	0	12
Forced type II	86	0	7

Forced type I = previously displaying birds forced into random pairs.
Forced type II = previously paired birds forced into random pairs. (After Bluhm, 1985)

pairings can have very serious consequences.

A lack of competition between males can apparently also have adverse effects on breeding success. In many primates which live in groups composed of several adult males and females, such as macaques, chimpanzees and most baboons, there is considerable competition between males for access to oestrous females in the group. When housed in groups which contain only one adult male, the lack of a competitor appears to reduce the motivation of the male to breed, and few young are produced (McGrew, 1981).

In addition to allowing animals to perform courtship displays and choose their mates, other social factors must be taken into account if captive breeding is to be successful and healthy populations maintained in zoos. As already discussed, it is important that the normal group structure is maintained, and that movements of animals between groups mimic the natural dispersal patterns. For some animals, other conspecifics are an important social stimulus in the breeding attempt, and without this social stimulation, successful breeding may not occur. In the kittiwake (*Rissa tridactyla*), a colonial breeding seabird, isolated pairs do not seem to be able to breed. In the dense colonies in which these birds normally breed, in addition to responding to the displays of their mates, breeding pairs also respond to displays of neighbouring birds. The social stimulation provided by other colony members seems to be essential for successful reproduction (Coulson, 1971). Such social facilitation is also important in some wildfowl (Bluhm, 1988) and there may be a minimum group size below which breeding will not take place.

A further problem in breeding animals in captivity is the avoidance of inbreeding. There is now substantial evidence of the deleterious effects which can result from matings between closely related individuals, the most obvious of these being high juvenile mortality (Ballou and Ralls, 1982; O'Brien *et al.*, 1985; Charlesworth and Charlesworth, 1987). A striking example is cheetahs (*Acinonyx jubatus jubatus*), which have great difficulty breeding successfully in captivity. Inbreeding amongst captive animals has been found to be so intense that skin grafts between most zoo cheetahs are not rejected, indicating the closeness of their genetic relatedness (O'Brien *et al.*, 1985). There is increasing evidence that animals ranging from arthropods to mammals may be able to recognize their close kin, even without prior contact (Kareem and Barnard, 1982). For example, infant macaques (*Macaca nemestrina*) have been shown to recognize half sibs even when they have been reared apart (McGrew, 1981). The adverse effects of inbreeding explain why many animals avoid mating with close relatives. This is especially well documented in chimpanzees, which avoid son-mother, father-daughter and sister-brother matings (Tutin, 1980). It has been suggested (Wright, 1932; Bateson, 1978b) that there is an optimum level of inbreeding which balances both the costs of inbreeding (the phenotypic expression of deleterious recessive genes through increased homozygosity) and distant outbreeding (the break-up of favourable gene complexes). A striking example of the effect of the latter is the sad case of the Tatra ibex (*Capra ibex ibex*). This ibex became extinct in Czechoslovakia, but was successfully re-established from Austria. Subsequently, two other sub-species were introduced from the Middle East and added to the Czech herd. The resulting hybrids rutted in early autumn rather than in winter as the local ibex did, which had the unfortunate consequence that the young were born in February, the coldest month of the year. The young did not survive and the whole Czech herd became extinct (Read and Harvey, 1986).

There are several behavioural mechanisms that potentially allow animals in the wild to optimize the level of inbreeding. In the majority of species for which there are adequate data, one sex disperses further than the other, thus reducing the likelihood of incestuous matings. Kin-recognition may allow animals to select mates of approximately the optimal degree of relatedness. For example Japanese quail (*Cotournix cotournix japonica*) prefer to associate with first cousins, apparently on the basis of plumage patterns (Bateson, 1982). Female great tits (*Parus major*) tend to mate with males which sing song types similar to, but not the same as, their fathers' (McGregor and Krebs, 1982). However, it is likely that in many species individuals estimate kinship on the basis of early experience, and so will avoid mating with individuals with whom they have been reared, regardless of genetic relatedness. The mechanisms of sexual imprinting may be important here, and the effects of early social experience, and there is likely to be an important sensitive period in early development when the effects of sexual imprinting occur (Bateson, 1978a,b). Thus zoos must exchange animals in order to overcome these constraints, and in attempts to conserve highly endangered species by captive breeding, great care must be taken to prevent the over-familiarization of young animals that could reduce the likelihood of their breeding together in future.

Maximizing the reproductive rate of commercially reared species must also take into account the likely life-history strategies to which their social behaviour is adapted. While many animals continue to grow throughout life, the growth rate is not constant. Sexual maturation tends to reduce, and often stop, growth, since resources are converted into gonadal rather than somatic tissue (Calow, 1981). This has important implications for commercial yields, as demonstrated in Chapter 5 for the salmonids. In addition, animals that have a life expectancy of more than one breeding season, may balance the costs and likely success of any one breeding attempt against the probability of their being able to breed successfully in the future. Thus an individual which is likely to breed several more times may abandon its current reproductive attempt if conditions appear unfavourable. The decision whether to continue should be dependent on the trade-off between the likely outcome of this attempt and its effect on future reproduction (Horn and Rubenstein, 1984). Decisions as to whether to breed or not are thus likely to be made on the basis of these physiological, environmental and social conditions which affect the costs of breeding. A greater understanding of these parameters should lead to our being able to modify the costs of breeding and so manipulate the decision in our favour. In addition to improving the productivity of commercial species, it may also prove possible to increase the reproductive output of endangered species and, by increasing the perceived costs of breeding, reduce the reproductive output of pests. These are tasks for the future!

REFERENCES

Avery, M.I. and Ridley, M.W. (1988) Gamebird mating systems, in *Ecology and Management of Gamebirds* (eds P.J. Hudson and M.R.W. Rands), BSP Professional Books, Oxford, pp. 159–76.
Baker, R.R. (1978) *Animal Migration*, Hodder and Stoughton, London.
Ballou, J. and Ralls, K. (1982) Inbreeding and juvenile mortality in

small populations of ungulates: A detailed analysis. *Biol. Conserv.*, **24**, 239–72.

Bateson, P.P.G. (1978a) How does behaviour develop? in *Perspectives in Ethology 3* (eds P.P.G. Bateson and P.H. Klopfer), Plenum Press, New York, pp. 55–66.

Bateson, P.P.G. (1978b) Sexual imprinting and optimal outbreeding. *Nature*, **273**, 659–60.

Bateson, P.P.G. (1982) Preferences for cousins in Japanese quail. *Nature*, **295**, 236–7.

Bertram, B.C.R. (1978) Living in groups: predators and prey, in *Behavioural Ecology: An Evolutionary Approach* (eds J.R. Krebs and N.B. Davies), Blackwell Scientific Publications, Oxford, pp. 64–96.

Bluhm, C.K. (1985) Male preferences and mating patterns of canvasback ducks (*Athya valisineria*), in *Avian Monogamy* (eds P.A. Gowaty and D. Mock), American Ornithologists Union Monographs.

Bluhm, C.K. (1988) Temporal patterns of pair formation and reproduction in annual cycles and associated endocrinology in waterfowl, in *Current Ornithology*, Vol. 5 (ed. R.F. Johnston), Plenum Press, New York.

Boyce, M.S. (1981) Beaver life-history responses to exploitation. *J. Appl. Ecol.*, **18**, 749–53.

Calow, P. (1981) Resource utilization and reproduction, in *Physiological Ecology: An Evolutionary Approach to Resource Use* (eds C.R. Townsend and P. Calow), Blackwell Scientific Publications, Oxford, pp. 245–73.

Chabrzyk, G. and Coulson, J.C. (1976) Survival and recruitment in the herring gull *Larus argentatus*. *J. Anim. Ecol.*, **45**, 187–203.

Charlesworth, D. and Charlesworth, B. (1987) Inbreeding depression and its evolutionary consequences. *Ann. Rev. Ecol. Systematics*, **18**, 237–68.

Clark, C.B. (1977) A preliminary report on weaning among chimpanzees of the Gombe National Park, Tanzania, in *Primate Bio-Social Development* (eds S. Chevalier-Skolnikoff and F.E. Poirier), Garland Press, New York, pp. 235–60.

Coulson, J.C. (1971) Competition for breeding sites causing segregation and reduced young production in colonial animals. *Proc. Adv. Study Inst. Dynamic Numbers Population* (Oosterkeek, 1971), 257–68.

Coulson, J.C. (In press) The population dynamics of culling herring gulls *Larus argentatus* and lesser black-backed gulls *L. fascus* in (eds C.M. Perrins, J.P. Lebreton and J.M. Hirones), *Bird Population Studies: relevance to conservation and management*, Oxford University Press, Oxford.

Coulson, J.C., Duncan, N. and Thomas, C.S. (1982) Changes in the breeding biology of the herring gull (*Larus argentatus*) induced by reduction in the size and density of the colony. *J. Anim. Ecol.*, **51**, 739–56.

Davies, N.B. and Houston, A.I. (1984) Territory economics, in *Behavioural Ecology* (eds J.R. Krebs and N.B. Davies), 2nd edn., Blackwell Scientific Publications, Oxford, pp. 148–69.

Feare, C. (1984) *The Starling*, Oxford University Press, Oxford.

Frazer, D. and Waddell, M.S. (1974) The importance of social and self grooming for the control of ectoparasitic mites on normal and dystrophic laboratory mice. *Laboratory Practice*, February 1974.

Gese, E.M., Rongstad, O.J. and Mytton, W.R. (1988) Home range and habitat use of coyotes in southeastern Colorado. *J. Wildl. Manage.*, **52**, 640–6.

Greenwood, P.J. (1980) Mating systems, philopatry and dispersal in birds and mammals. *Anim. Behav.*, **28**, 1140–62.

Greig, S.A., Coulson, J.C. and Monaghan, P. (1985) Feeding strategies of male and female adult herring gulls (*Larus argentatus*). *Behaviour*, **94**, 41–59.

Harris, S. and Trewhella, W.J. (1988) An analysis of some of the factors affecting dispersal in an urban fox (*Vulpes vulpes*) population. *J. Appl. Ecol.*, **25**, 409–22.

Hasegawa, T. and Hiraiwa, M. (1980) Social interactions of orphans observed in a free range troop of Japanese monkeys. *Folio primatologica*, **33**, 129–58.

Holst, D. von (1974) Social stress in treeshrew: causes and physiological and ethological consequences, in *Prosmian Biology* (eds R.D. Martin, G.A. Dayleit and A.C. Walker), Duckworth, London.

Horn, H.S. and Rubenstein, D.I. (1984) Behavioural adaptations and life history, in *Behavioural Ecology: An evolutionary approach* (eds J.R. Krebs and N.B. Davies), 2nd edn., Blackwell Scientific Publications, Oxford, pp. 279–98.

Hudson, P.J. and Rands, M.R.W. (eds) (1988) *Ecology and Management of Gamebirds*, BSP Professional Books, Oxford.

Johnston, W.E. (1965) On mechanisms of self-regulation of population abundance in *Oncorhynchus nertea*. *Mitt. Int. Ver. Theor. Ange. Limnol.*, **13**, 66–87.

Kareem, A.M. and Barnard, C.J. (1982) The importance of kinship and familiarity in social interactions between mice. *Anim. Behav.*, **30**, 594–601.

Kenward, R.E. (1978) The influence of human and goshawk *Accipiter gentilis* activity on woodpigeons *Columba palumbus* at brassica feeding sites. *Ann. Appl. Biol.*, **89**, 277–86.

Kleiman, D.G. (1980) The sociobiology of captive propagation, in *Conservation Biology* (eds M.E. Soule and B.A. Wilcox), Sinauer, Sunderland, Mass., pp. 234–61.

Krebs, J. and Davies, N.B. (eds) (1984) *An Introduction to Behavioural Ecology*, 2nd edn., Blackwell Scientific Publications, Oxford.

Kruijt, J.P. and Hogan, J.A. (1967) Social behaviour on the lek in black grouse *Lyrurus tetrix tetrix, Ardea*, **55**, 203–240.

Kurland, J.A. (1977) Kin selection in the Japanese monkey. *Contributions to Primatology*, **12**, 1–145.

Lawrence, A.B. and Wood-Gush, D.G.M. (1988) Influence of social behaviour on utilization of supplemental feed blocks by Scottish hill sheep. *Anim. Prod.*, **46**, 203–12.

MacDonald, P.W. (1980) *Wildlife and Rabies*, Oxford University Press, Oxford.

McArthur, K.L. (1981) Factors contributing to the effectiveness of black bear transplants. *J. Wildl. Manage.*, **45**, 102–10.

McGregor, P.K. and Krebs, J.R. (1982) Mating and song types in the great tit. *Nature*, **297**, 60–1.

McGregor, W.C. (1981) Social and cognitive capabilities of non-human primates: lessons from the wild to captivity. *Int. J. Stud. Anim. Prob.*, **2**, 138–49.

Monaghan, P. (1989) Communal roosting and social behaviour of choughs. *Proc. International Chough Workshop 1988*, Nature Conservancy Council, UK.

Monaghan, P., Shedden, C.B., Ensor, K. *et al.*, (1985) Salmonella carriage by herring gulls in the Clyde area of Scotland in relation to their feeding ecology. *J. Appl. Ecol.*, **22**, 669–80.

Moss, R. and Watson, A. (1984) Adaptive value of spacing behaviour in population cycles of red grouse and other animals, in *Behavioural Ecology* (eds R.M. Sibly and R.H. Smith), Blackwell Scientific Publications, Oxford, pp. 273–94.

Murphy, G.I. (1980) Schooling and the ecology and management of marine fish, in *Fish Behaviour and Its Use in the Capture and Culture of Fishes*, (eds J.E. Bardach, J.J. Magnusson, R.C. May and J.M. Reinhart), ICLARM Conference Proceedings, Vol. 5, ICLARM, Manila, Philippines, pp. 400–14.

Neill, S.R.St.J. and Cullen, J.M. (1974) Experiments on whether schooling by their prey affects the hunting behaviour of cephalopods and fish predators. *J. Zool.*, **172**, 549–69.

Nisbet, I.C.T. (1973) Courtship feeding, egg size and breeding success in Common Terns. *Nature*, **241**, 141–3.

O'Brien, S.J., Roelke, M.E., Marker, L. *et al.*, (1985) Genetic basis for species vulnerability in the cheetah. *Science*, **227**, 1428–34.

Packer, C. (1975) Mate transfer in olive baboons. *Nature*, **255**, 219–20.

Partridge, L. (1980) Mate choice increases a component of offspring fitness in fruit flies. *Nature*, **283**, 290-1.

Partridge, L. and Halliday, T.M. (1984) Mating patterns and mate choice, in *Behavioural Ecology: An Evolutionary Approach* (eds J.R. Krebs and N.B. Davies), 2nd edn., Blackwell Scientific Publications, Oxford, pp. 222-50.

Pulliam, H.R. and Caraco, T. (1984) Living in groups: Is there an optimal group size?, in *Behavioural Ecology: An Evolutionary Approach* (eds J.R. Krebs and N.B. Davies), 2nd edn., Blackwell Scientific Publications, Oxford, pp. 122-47.

Pusey, A.E. (1980) Inbreeding avoidance in chimpanzees. *Anim. Behav.*, **28**, 543-52.

Read, A.F. and Harvey, P.H. (1986) Genetic management in zoos. *Nature*, **322**, 408-10.

Robertson, P.A. and Rosenberg, A.A. (1988) Harvesting gamebirds, in *Ecology and Management of Gamebirds* (eds P.J. Hudson and M.R.W. Rands), BSP Professional Books, Oxford.

Slaney, P.A. and Northcote, T.G. (1974) Effects of prey abundance on density and territorial behaviour of young rainbow trout (*Salmo gairdneri*) in laboratory stream channels. *J. Fish Res. Board Can.*, **31**, 1201-9.

Still, E., Monaghan, P. and Bignal, E. (1986) Social structuring at a communal roost of choughs. *Ibis*, **129**, 398-403.

Stolba, A. (1985) Minimizing social interference during feeding in pig groups, in *Abstracts of the 19th International Ethological Conference, Vol. 2*, Toulouse, France, p. 460.

Ten Cate, C. (1989) Behavioural development: toward understanding processes, in *Perspectives in Ethology 8* (eds P.P.G. Bateson and P.H. Kopfler), Plenum Press, New York, pp. 243-69.

Till, J.A. and Knowlton, F.F. (1983) Efficacy of denning in alleviating coyote depredations upon domestic sheep. *J. Wildl. Manage.*, **47**, 1018-25.

Trewhella, W.J., Harris, S. and McAllister, F.R. (1988) Dispersal distance, home range size and population density in the red fox (*Vulpes vulpes*): a quantitative analysis. *J. Anim. Ecol.*, **25**, 423-34.

Tutin, C.E.G. (1980) Reproductive behaviour of wild chimpanzees in the Gombe National Park, Tanzania. *J. Reprod. Fertility Suppl.*, **28**, 43-57.

Windberg, L.A. and Knowlton, F.F. (1988) Management implications of coyote spacing pattern in southern Texas. *J. Wildl. Manage.*, **52(4)**, 632-40.

Wright, S. (1932) The roles of mutation, inbreeding, crossbreeding and selection in evolution. *Proc. 6th International Congress of Genetics*, 356-66.

4

Communication

IAN R. INGLIS AND DAVID S. SHEPHERD

"But animals don't always speak with their mouths", said the
parrot in a high voice, raising her eyebrows. "They talk with their
ears, with their feet, with their tails — with everything ... The
world has been going on now for thousands of years hasn't it?
And the only thing in animal-language that *people* have learned
to understand is that when a dog wags his tail he means 'I'm
glad'! — it's funny isn't it? You are the very first man to talk like
us. Oh, sometimes people annoy me dreadfully — such airs they
put on talking about 'the dumb animals'. *Dumb*!"

From *Dr Doolittle* by Hugh Lofting.

The majority of the species that man attempts to manage, whether
in agriculture, wildlife conservation or pest control, live in social
groups. Coordinated interactions between individuals form the basis
of social life and communication between animals is essential for
such interactions. The essence of animal communication is that
one animal (the 'sender') behaves in a manner which influences the
behaviour of other animals (the 'receivers'). As we shall discuss
below, riders have to be placed on this bald statement. For example,
an animal's behaviour can be influenced by physical force but such
actions are not usually considered to be part of communication; as
Cherry (1957) argued, to a man the command 'Go jump in the pool' is
a signal, but the push which precipitates him is not. This example illustrates another aspect of communication. The shouted command,
assuming it is obeyed, requires far less energy from the sender to
achieve the end-point of a person in the water than would a push.
Communication is typically very energy efficient. The male firefly does
not actively search for a female but emits a series of flashes which
attract females to him. If, therefore, we wish to make an animal more,
or less, likely to behave in a certain way, an efficient approach is to
utilize the signals other animals use to gain the same ends. Before
tackling the various ways in which this has been achieved, or

attempted, we begin with a brief introduction to the study of animal communication. There is a vast literature on this subject and extensive reviews can be found for example in Sebeok (1968, 1977), Smith (1977), Green and Marler (1979), and Halliday and Slater (1983), whilst a good discussion of the underlying principles is provided by Cullen (1972).

4.1   THEORY

4.1.1 The nature of animal communication

An animal's behaviour can be influenced in many ways. As already mentioned, changing another's behaviour by physical force is not considered as communication. What about the mouse rustling in the leaves: is it communicating to the owl? Marler (1967) argued not, since the sender clearly does not benefit from the transmission of the sounds. 'For an example of communication to have evolved it is essential that signal transmission should, on average, be advantageous to the sender' (Slater, 1983). Natural selection results in average statistical benefits so that although there may be a few circumstances where a signal is clearly disadvantageous, nevertheless the communication will normally be well adapted to the 'usual' circumstances. It can however be difficult to demonstrate an advantage even under the normal circumstances in which a signal is used. For example, there are a plethora of suggestions (Harvey and Greenwood, 1978) as to why such apparently altruistic behaviour as giving an alarm call does in fact benefit the sender, yet little in the way of experimental evidence (Møller, 1988). Nevertheless, by laying stress upon the functional advantage of the communication to the sender it is possible to avoid definitions based upon the intention to communicate (MacKay, 1972). This is a valid approach in studies of human communication (Goffman, 1969) when the subjects can be questioned on their intentions but it is hard to imagine how such information could be obtained from other animals.

If the sender is thought invariably to benefit from the communication, does the same hold true for the receiver? This is not the case for most examples of interspecific communication where, as with the gaping of the cuckoo nestling, the receiver certainly does not benefit. In such instances the sender deceives the receiver by mimicking a signal to which the receiver has been selected to respond. Feeding gaping nestlings does on average benefit the cuckoo's host since those nestlings are usually its own. However, the receiver need not always lose out. For example, many species of passerine birds forage in mixed-species groups. Individuals in such groups respond to the alarm calls given by members of the other species since, in the presence of a predator, the receiver benefits from the maintenance of the flock (Lazarus, 1972) as well as the sender.

Dawkins and Krebs (1978) have argued that the manipulation of the receiver by the sender is also the norm in intraspecific communication. Their position stems from the work of Maynard Smith (1974), who used a game theory approach to argue that individuals would not be expected to provide truthful information when communicating, or to respond to other's signals as if they were truthful, because this strategy could be exploited by cheaters. The idea that deception lies at the heart of all communications is in stark contrast with the view of other workers (Cullen, 1966; Smith, 1977) who maintain the sender provides accurate information to conspecifics. There has been much debate on this issue (Hinde, 1981; Caryl, 1982; Enquist, 1985) and the true position would appear to lie between the extremes. Whilst there are clear examples of deceit within intraspecific communication (Dominey, 1981), to suggest that such manipulation is the basis of all communication is a gross over-emphasis.

Many signals simply cannot be faked, Antler size and form in ungulates is one example (Geist, 1971). The depth of the pitch of aggressive calls in toads (*Bufo bufo*) is related to their body size which again cannot be faked (Davies and Halliday, 1978). Olfactory cues can form part of the signal and be impossible to fake, as in squirrel monkeys (*Saimiri sciureus*) where the communication of dominance involves penile erection and urination. This display is related to hormonal levels (Ploog *et al.*, 1963) and the olfactory cue in the urine is necessarily an accurate indicator of the sender's physiological state (Baldwin, 1968). Signals are often very repetitive and it has been found in many species (e.g. iguanas (*Iguana iguana*), Rand and Rand, 1976; red grouse (*Lagopus lagopus*), Watson and Miller, 1971; red deer (*Cervus elaphus*), Clutton-Brock and Albon, 1979) that the frequencies and/or durations of the signals are related to the sender's dominance. Such signals must be well correlated with the general body condition of the sender; the stamina needed for these contests cannot be faked.

Even if the signals could be faked there are situations in which deception would not result in any advantage to the sender. For example, giving false signals of high rank can result in greater attack by truly dominant individuals (Rohwer, 1977). Furthermore in situations where the deceiver can expect some short-term advantage this can be outweighed by a longer-term penalty. There is an advantage in sharing accurate information with other members of a social group in that this allows the efficient integration of their behaviour when combatting predators, foraging, etc. A similar level of efficiency would not be possible in a society of liars since the reliability of any communication would be too low. The ability to recognize and remember many individuals (Falls, 1982) may have evolved to enable receivers to downgrade the reliability they place upon communication from senders who have deceived them in the past; thus vervet monkeys (*Cercopithecus aethiops*) learn to ignore calls given by unreliable individuals (Cheney and Seyfarth, 1988). Any success of the cheat will thereby be short-lived and, in such societies, there will be selection for truthful signals (Van Rhijn and Vodegel, 1980). As Hinde (1981) states 'the general utility of terms like "deceit" and "manipulation" for animal signalling systems remains to be established'.

The term 'influence' in our definition clearly emphasizes that communication is probabilistic; the receiver becomes more or less likely to perform a certain behaviour. Communication does not trigger a response automatically and an animal's experience can greatly influence how it responds to a signal: for these reasons pheromones will not be considered in this chapter. Pheromones are 'isolated chemicals shown to be relatively species-specific which elicit a clear and obvious behavioural or endocrinological function and which produce effects involving a huge degree of genetic programming and influenced little by experience' (Martin, 1980). There is a distinction between the way in which insect behaviour can be repeatedly triggered by exposure to pheromones and the flexible response shown by vertebrates to their chemical signals. As Bronson (1976) stated 'it is difficult to imagine a male mouse attempting to copulate with a piece of scented filter paper let alone doing it repeatedly'. Figure 4.1 summarizes the main aspects of the interchange of information that is considered to be communication.

4.1.2 Signals

The information passed between communicating animals is encoded within signals. A distinction has to be made (Smith, 1968) between the message of the signal and its meaning. Slater (1983) provides a good example of the way these terms are used. During the breeding season the unmated males of many songbird species sing the most. The message of this song would therefore be 'here is an unmated male'. The meaning of the song would however vary with the recipient. To another male of the same species it would signal an occupied territory; to the female of the species the meaning of the song would be

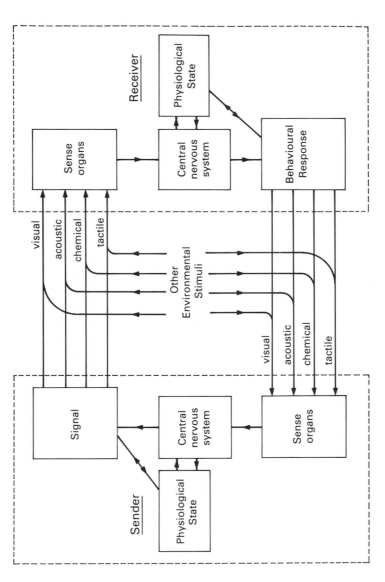

Figure 4.1 Diagram of the main features involved in communication. The sender emits a signal whose message is influenced by the physiological state of the animal as well as by the actions of the central nervous system. The signal becomes embedded in environmental noise and the resulting complex stimulates the sense organs of the receiver. The receiver's central nervous system attaches a meaning to the perceived signal and this meaning directly affects both the physiological state of the receiver and the type of behavioural response made, which may be another signal.

'there is a possible mate', whilst to a predator it would signify the presence of a potential meal. The meaning of a signal not only varies between recipients but also with the context in which the signal is used. A song of a neighbouring bird from within its own territory will not elicit attack from the adjacent territory owner but the same song· sung from an unexpected location will (Falls and Brooks, 1975). Indeed some signals are used to change the context, and hence the meaning, of others; as in primates where the 'play face' signifies that the accompanying boisterous behaviour is not real fighting (Andrew, 1963). Furthermore, as already mentioned, the meaning may vary with the identity of the sender (Cheney and Seyfarth, 1988). In such ways subtlety of communication can be achieved with a relatively low number of signals.

It can be difficult to decide whether a particular behaviour acts as a signal since we must determine whether it causes a shift in the likelihood that the receiver will show a particular response. The difficulty is greatest when there is a long gap between signal and response. For example Lehrman and others (Lehrman and Friedman, 1969) have shown that visual and auditory signals involved in courtship can affect the reproductive behaviour shown by the female ring dove (*Streptopelia capicola*) much later.

Even when the interval between signal and response is short there is always the danger of confusing the correlation of two events with true causation. Just because one event reliably precedes another it does not necessarily mean that the first causes the second; they could both be controlled by a common earlier event. The only satisfactory way to resolve this problem is to manipulate the behaviour of one of the participants experimentally using models or recorded sounds, but this is not always possible. Furthermore, before attempting any experiments it is necessary critically to observe natural episodes involving the signals concerned since only by this means can relevant experimental hypotheses be formulated. During such observations we can look at the sender and ask what are the likely causes of the signal; in this way intelligent guesses can be made as to its message. We can also measure the changes in the behaviour of the receiver in order to discover the meaning of the signal. Both aspects involve the analysis of sequences of behaviour which is itself not without difficulties (Slater, 1973).

The nature of the signal will vary with the quality of the information it encodes. If, for example, the message is simply the species identity of the sender then the signal can be uncomplicated and stereotyped. If however it conveys detailed information, such as 'here is a mated male goose defending a nesting territory with an increasing tendency to attack', then the signals involved will be variable both in form and arrangement. The more variable the signal, however, the greater the risk that the receiver will either not recognize it as such or derive from it a meaning very different from the intended message. Under some environmental conditions (loud background noise, low light levels), it may be impossible for the receiver to discriminate between one variant of the signal and another.

The basic problem is therefore to produce a signal that minimizes ambiguity so that the perceived meaning corresponds well to the conveyed message. It is important to distinguish between this type of ambiguity and ambiguity which may be a fundamental part of the actual message. For example Caryl (1979) reanalysed data on avian threat signals (Stokes, 1962; Dunham, 1966) and showed that they were associated only with probabilities of subsequent actions by the sender rather than precise predictions. Hinde (1981) has pursued such analysis further and concludes that the message in such displays is of the type 'I want to stay but if you do X I am more likely to escape than to attack you' where X could approach, threaten, etc. More recent studies (Piersma and Veen, 1988) have also revealed the existence of truthful but conditional threat displays. The message itself is ambiguous since the subsequent behaviour of the sender depends upon the response of the receiver. The behaviours used to encode this ambiguous message, however, have been selected for minimal uncertainty.

4.1.3 Ritualization and displays

Behaviours that have evolved to function as signals are called displays and the process by which they have been shaped to fulfil this role is called ritualization (Tinbergen, 1952). If an animal can benefit by predicting the behaviour of another then it will do so. A goose has to extend its neck and open its beak before pecking; a dog has to bare it teeth before biting. Individuals that are able to recognize such preparatory movements can retreat and thereby avoid being pecked or bitten. Similarly if a goose or dog can cause conspecifics to flee simply by extending a neck or exposing teeth, selection will enhance this ability by making such movements more conspicuous. The behaviour becomes ritualized. Ritualization involves two major processes that operate in parallel. The first involves changes in the behaviour itself. A display can differ from the original behaviour in a number of ways: in frequency of occurrence, intensity and speed of performance, orientation, the nature of coordination between components of the behaviour, the loss of some of these components, and in the development of rhythmic repetition (Morris, 1957; Hinde, 1959). The second major process is the development of structures (e.g. crests, plumes, coloured patches of skin) and/or calls that greatly enhance the conspicuousness of the behaviour.

If there are good reasons to believe that a number of displays have a common ancestral behaviour then comparisons between species of the form of their displays can yield valuable information (Baerends, 1958; but see also Atz, 1970). The comparative study of displays led to the early realization (Daanje, 1950; Tinbergen, 1952, 1959) that displays evolve from three major classes of behaviour. Many originate from automatic responses such as movements of hair, urination, defaecation and changes in skin colour brought about by dilation and constriction of surface blood vessels (Morris, 1956). Incomplete preparatory movements, or 'intention movements' (Tinbergen, 1952), seen at the beginning of an activity, are perhaps the most common behaviours that have been ritualized. These intention movements need not be associated with aggressive acts. Intention movements associated with preening, copulation, nesting and many other behaviours have also been ritualized into displays (Andrew, 1961; Moynihan, 1955). The last class is more difficult to explain, not least because there appears to be no single causal factor associated with these behaviours; these have been called 'displacement activities' (Tinbergen, 1952; Bastock *et al.*, 1953). 'A displacement activity is an activity belonging to the executive motor pattern of an instinct other than the instinct(s) activated' (Tinbergen, 1952). For example male birds often break off courting a female in order to indulge in mock preening or drinking; behaviours which appear to be totally irrelevant to the task in hand.

There have been many theories as to the motivation underlying displacement behaviour (Hinde, 1970), but once such activities have evolved into displays this question may no longer be relevant. It has been suggested (Tinbergen, 1952; Blest, 1961) that during the course of ritualization the display becomes freed, or 'emancipated', from the causal factors that mediated the original behaviour. In the courtship displays of many ducks, displacement preening and drinking have been ritualized into a stereotyped pattern (Lorenz, 1961). Gannets and boobies (Sulidae) have a highly specialized 'sky-pointing' display which seemingly has evolved from a pre-flight posture but now is used for sexual advertisement (Nelson, 1969). Courtship feeding in birds and genital presentation in primates are other examples of displays apparently now being used to encode information about a motivational state different from that which mediated the ancestral behaviour. However, as Hinde (1970) argued, it is difficult to establish just how much emancipation has occurred since it cannot be proved that the causal factors associated with the display were not in fact also in some way involved with the original behaviour.

Displays are far more stereotyped than the original behaviour. For example the 'head-throw' display of the Goldeneye duck (*Bucephala clangula*) has a mean duration of 1.29 seconds with a standard error of plus or minus only 0.08 seconds (Dane *et al.*, 1959). The distinct and rigid nature of many displays led early workers (Tinbergen, 1942) to call them Fixed Action Patterns. It was further argued by Morris (1957) that displays are performed at a 'typical intensity' in that the form of display varies little in response to changes in the level of the causal factors associated with it. It is unlikely that an animal edging towards a rival has the same combination of fear and aggression as an animal edging away and yet the same threat posture can be shown in both circumstances. Morris (1957) argued that 'the reduction in the amount of information provided by the signal about the exact state of the signaller that is involved with this stereotyping, is apparently more than compensated for by the elimination of signal ambiguity. A signal that is constant in form cannot be mistaken.' Proponents of the 'manipulation theory' of communication argue that the loss of information concerning the 'state of the signaller' is what stereotyping and typical intensity are all about. It is not so much to make the signal more recognizable but to conceal the likely behaviour of the sender (Maynard Smith, 1972). However it is difficult to explain on this basis why signals conveying apparently opposite meanings are also so often opposite in form. Darwin first drew attention to this 'principle of antitheses' by pointing to the extreme difference in posture between a hostile dog and one 'in a humble and affectionate frame of mind', The former has the head and neck thrust forward, the ears pricked, raised hair on the neck and back and an upright and rigid tail. The latter by contrast has the head and neck arched upwards, the ears flat, the hair sleek and the tail lowered and wagging (see Figure 4.2). Similarly, in gulls the beak is made conspicuous in aggressive displays but hidden, by turning the head away, in appeasing ones (Tinbergen, 1959). Such contrasts are more easily explained by the need to reduce any chance that the displays could be confused. It should also be remembered that when a stereotyped signal is used as a repetitive unit in a prolonged display, a necessarily accurate indication of the prowess of the sender is conveyed by the frequency of the signal and the duration of the display.

More recent research has revealed that many displays are not quite as fixed as previously thought. Rather there is some variation around a most common form and this has led Barlow (1977) to propose that the term 'Fixed Action Pattern' should be changed to 'Modal Action

Figure 4.2 Behavioural signals from aggressive and submissive dogs. (Modified from Darwin, 1872.)

Pattern'. It is not known whether such variation does itself encode additional information.

4.1.4 Constraints on signal design

Each of the various media through which information can be transmitted has its own advantages and disadvantages. Chemical communication is the most widespread and was probably the first to evolve. Chemical signals can be released directly into the environment by opening or everting storage glands or secretory surfaces on the skin. They can also indirectly pass to the exterior via urine and faeces. The major problem with chemical communication is that in still air or water, dispersion occurs very slowly by diffusion. After one second from emission in still air only 10% of a typical pheromone has managed to travel as far as 1cm from the source (Shorey, 1976). One solution to this problem is to send the signal into a wind or current but then the direction of the transmission is restricted. Furthermore turbulence will disperse the molecules of the signal and so decrease the downstream reception distance. Insects have greatly increased the power of their receptors to compensate and many can be triggered by a single molecule of pheromone (Boeckh *et al.*, 1965). Although chemical signals are usually most effective over relatively short distances the effective range is tailored to suit the type of message conveyed.

Individual identification odours need only to operate at very close range and hence are frequently relatively non-volatile, large molecular weight compounds, such as the glycoproteins which distinguish individual rats (Singh *et al.*, 1987). Territorial scent marks should be more volatile but sufficiently long-lived to deter intruders until the resident can re-mark the spot; the anal gland secretions of the mongoose (*Herpestes auropunctatus*) function in this way and contain a mixture of six, short-chain carboxylic acids (Gorman, 1976). Chemical cocktails are also to be found in signals of sexual condition, such as rat preputial gland secretions which contain various aliphatic acetates to which male and female rats react differently (Stacewicz-Sapuntzakis and Gawienowksi, 1977). Such mixtures include short chain (two to five carbons), volatile compounds to advertise and longer chain (up to 20 carbons), less volatile compounds that facilitate location of mates by following trails. Alarm signals on the other hand should be non-directional and most effective when detected at distance. They need to be highly volatile. The ischiadic gland secretion of the pronghorn (*Antilocarppa americana*) is released when the animal is startled and, given favourable wind conditions, can apparently be detected by another individual up to a mile away (Seton, 1927).

Vertebrates, in particular the mammals, exploit one of the major advantages of chemical communication; that the signal may be active for long periods in the absence of the sender. The use of secretions to mark territories and home ranges is widespread. For example the European rabbit (*Oryctolagus cuniculus*) marks its territory by 'chinning' objects with secretions from submandibular glands (Mykytowycz, 1965); alligators (*Allegator* sp.) mark breeding territories with secretions from an anal gland (Lederer, 1950). The ability to divorce the signal from the sender means there is less chance that a predator will be able to locate the source of the signal. However, it also means that, in comparison with visual and acoustic signals, chemical communication can only convey crude temporal information.

The major advantages of visual communication are its speed and directionality. However visual communication depends upon ambient light except for those few animals that can generate their own light. Therefore, unlike chemical and acoustic signals, the intensity of visual signals cannot normally be varied above a maximum set by the ambient light level. This is a severe restriction in certain habitats (e.g. caves, muddy water, dense vegetation), and at night, for even bright moonlight is 10 log units less intense than daylight. Brightness can be adjusted to some degree by the use of reflecting surfaces; thus by performing its display dive in relation to the sun the Anna's hummingbird (*Calypto anna*) ensures the receiver sees a brilliant reflection from irridescent head feathers (Hamilton, 1965). Most animals, however, use pattern, tone and colour to increase the conspicuousness of their visual signals within the limitations imposed by the ambient light level. Visual communication on land is constrained by the scattering of light which reduces contrasts in tone and colour. The shorter wavelengths are scattered the most and diurnal animals often have a yellow filter in front of the receptor surface which reduces the transmission of these wavelengths and also cuts the light scatter within the eye (Tansley, 1965). Nocturnal animals lack such filters. Water is generally a turbid medium, which absorbs especially the longer wavelengths, thereby restricting the range of visual signals. Any attempt in water or air to increase the effective distance of a signal by increasing its conspicuousness can lead to higher predation risk. Many visual signals are therefore turned on and off by movements that reveal and then conceal them; these actions also allow control of the timing of transmission of the information. Alternatively, visual signals which have to be continuously active for long periods (e.g. breeding plumage in ducks) may be shed as soon as they are no longer needed.

Unlike visual signals, sound travels well in both air and water and is not restricted by the ambient light level. Unlike chemical signals, the transmission of sound over large distances does not require a wind or current and can be directed in any direction. Auditory signals also fade quite rapidly in most habitats (the amplitude generally declines by about 6dB for each doubling of distance) and may therefore convey temporal information. However, sound production can be expensive in energy. Around three-quarters of the energy assimilated during the six month breeding season of two species of Australian frogs (*Ranidella* sp.) is used in singing (MacNally, 1981). A singing cicada (*Cystosoma saundersii*) can have a metabolic rate of 20 times that of the resting insect (MacNally and Young, 1981). Indeed it is the physical effort needed to make certain acoustic signals that, as we mentioned, makes them impossible to fake, as in the 'roaring' display of red deer (Clutton-Brock and Albon, 1979).

As the frequency of sound increases there is a greater attenuation both from atmospheric absorption and from scattering (Wiley and Richards, 1982). This relationship is found irrespective of habitat, although in forests the scattering is mainly caused by vegetation whilst in open terrain it stems from atmospheric turbulence. Complex boundary effects come into play when sound is propagated within 1–2m of the ground but, as a general rule, to reduce attenuation it is best to use the lowest frequency possible. The booming calls of grouse are heard several kilometres away and owls, doves and cuckoos use frequencies below 1kHz for long distance communication (Greenwalt, 1968). It is true that many small animals use higher frequencies but this seems to be a constraint of their body size. Small sound sources cannot generate low frequencies efficiently and there is a rough correlation between the dominant frequencies used in a bird's song and its size (Konishi, 1970). Some animals get around this problem by using a resonating horn; in this way a hollow tree enables low frequencies to be produced by even small woodpeckers. It should be remembered, however, that attenuation can be an advantage in limiting the spread of signals, as when communicating between mates or between parents and young.

Sounds in general are less easy to locate than visual signals and more so than chemical ones. Certain properties of the sound determine whether it is easy to locate (Marler, 1959). The source of sound consisting of a wide range of frequencies with high amplitude variations is not difficult to pinpoint since interaural comparisons are made easier. The distress scream emitted when a bird is seized by a predator is of this type. This signal usually elicits mobbing of the predator by conspecifics and hence accurate localization is crucial. By contrast the alarm call emitted when a bird sights a predator is of high frequency and short duration; properties which should make it difficult for the predator to locate (Brown, 1982).

Under natural conditions a receiver has to distinguish between a large number of signals from its own and other species, and be able to detect them within the general background 'noise'. Noise in this context refers to irrelevant stimuli regardless of modality. Wind in the trees, the sounds of running water, etc., are examples of acoustic noise. The number of organic molecules suitable for use as chemical signals is vast and hence chemical noise is mainly composed of odours from other species. In visual communication the scattering of light from vegetation, reflection from water, etc., form part of the background noise. Signals detection theory attempts to characterize the optimum properties of a receiver trying to detect signals embedded in noise (Egan, 1978). The main conclusion of the theory is that a receiver cannot increase the probability that it will correctly detect a signal without also increasing the probability that it will make a false alarm. However, it is possible to approach the optimal balance between these two aspects by improvements in signal design.

The performance of a receiver is dependent upon two factors; the inherent detectability of the signal in the noise, and the criterion set by the receiver for the presence of a signal. It has been shown that maximum performance is reached when the receiver knows in advance the precise form of the signal and when it will occur. As a result there are three major adaptations by which performance can be improved (Wiley and Richards, 1978, 1982); reduction in the diversity of alternative signals, repetition and alerted detection.

As we have already discussed, ritualization does result in a reduction in the variability of signals and furthermore the resultant displays are frequently repeated. In ecological communities with great species diversity we might also expect smaller repertoires of signals in order to improve recognition against the background of many heterospecific signals. This effect has been reported for North American wrens (*Troglodytes* sp., Kroodsma, 1977). The adaptation of alerted detection is only possible for signals that can convey accurate temporal information. It is necessary that some easily detectable signal shortly precedes in a predictable manner a second signal which carried the important information (Raisbeck, 1963). The first signal alerts the receiver to expect the second within a short time and this greatly increases the probability of detection. For example, experiments involving the playback of various sections of the song of the rufous-sided towhee (*Pipilo erythrophthallmus*, Richards, 1981) reveal that the initial note serves only to alert the receiver and that as a result recognition of the subsequent trill part of the song is enhanced. Many birds begin their territorial songs with a short series of simple tones that are easily detectable and then the song becomes progressively more complex (e.g. indigo bunting, *Passerina cyanea*, Shiovitz, 1975). Although the evidence for alerted detection involves acoustic communication there is no reason why acoustic signals could not work in conjunction with visual ones. The sound of the wingclaps made when a woodpigeon (*Columba palumbus*) flies may alert conspecifics and make it easier for them to detect a visual alarm signal associated with it (Inglis and Isaacson, 1984).

This account of animal communication has neglected a number of issues (e.g. the need for reproductive isolation between species can greatly influence the signals used in courtship; Hinde, 1959); however, the concepts most relevant to the applied use of signals have been briefly introduced and we turn to consider the practical uses of animal signals.

4.2   PRACTICE

4.2.1 The use of intraspecific signals

Almost all the natural functions of odour signals have some potential for exploitation. Long before there was any scientific interest in chemical communication, fur trappers were using extracts such as musk or beaver castor to attract quarry into traps (Bateman, 1973; Schemnitz, 1980). Such chemicals are used to mark territorial boundaries but the message of these signals must be more than just 'Keep out' otherwise the technique would be useless (Stoddart, 1980). In fact the receiver actively investigates the signal to obtain additional information on, for example, the number of individuals that have marked there or their sexual and social status.

The use of odours to identify individuals is particularly important for very young animals of species such as goats (Klopfer and Gamble, 1966), pigs (Meese *et al.*, 1975) and sheep (Bouisson, 1968) which must be recognized by the mother before being allowed to suckle. The mother learns the smell of her offspring within the first hours of its life. In order to get an orphan lamb accepted by a mother whose own offspring has recently died, it is common practice for shepherds to dress the orphan in the skin of the dead lamb, or to cover it with fleece and milk from the foster mother. In this way the unfamiliar smell of the orphan is masked by the familiar. The same technique was used to cross-foster a newborn male vicuna to an alpaca mother in order to facilitate hybridization of the two species and thereby produce a domesticated animal yielding vicuna wool in alpaca quantities (Hodge, 1946). More recently fostering beef calves has been facilitated by transferring odour from the mother's own offspring to the orphan via cloth jackets (Dunn *et al.*, 1987).

The lack of a familiar odour could be the major trigger of aggression and stress when domesticated animals such as pigs, sheep or cattle are introduced into existing groups. As our knowledge of chemical markers grows it may be possible to spray the newcomers with an extract of the 'colony' odour which will usually approximate the odour of the dominant animals for these engage in most marking behaviour. If synthetic odours are applied to very young animals the chances of natural odours being confounded with the synthetic ones are greatly increased. Mainardi *et al.*, (1965) found that females in litters of mice that had been scented with 'Violetta di Parma' perfume at 5 to 20 days of age, later preferred unperfumed males, indicating that they associated the perfume with siblings, who are not chosen as mates (Hepper, 1986). Rats have similarly been imprinted on a range of odours such as 'red rose cologne' (Marr and Gardner, 1965). Can we look forward to piglets literally smelling of violets?

Chemical signals play an important part in the sexual behaviour of many species either as attractants or as primers which trigger sexual receptivity. The effect of the odour of boar, supplied as an aerosol preparation, in eliciting lordosis in oestrous females (Melrose *et al.*, 1971) is a well known example of the use of such chemical signals in animal husbandry. Identification of females in breeding condition within groups in which oestrus is not synchronized can greatly enhance breeding efficiency. Oestrous gilts will themselves choose the odour of a boar whilst anoestrous females show no preference (Signoret, 1970). Cattle dogs can be trained to locate suitable heifers for insemination (Stoddart, 1980). In species such as sheep, in which oestrus can be induced, the ovulation of the flock can be synchronized by the introduction of a male shortly before the beginning of the mating period (Brown, 1979) with obvious advantages in the lambing season. It appears that a novel ram is effective whereas one that has been in contact with the ewes during the inactive period is not (Hulet, 1966). A similar 'priming' effect will accelerate puberty in the pig (Brooks and Cole, 1970), and shorten the time to first oestrus after parturition by as much as two weeks (Hillyer, 1976), which considerably increases profits.

Sexual signals have potential for use in other areas. The urine of pre-oestrous females is very alluring to adult males of many rodent species (Shumake, 1977). The active ingredients of rat preputial gland secretions have been isolated (Gawienowski *et al.*, 1975, 1976) and their attractive properties demonstrated (Stacewicz-Sapuntzakis and Gawienowski, 1977; Gawienowski and Stacewicz-Sapuntzakis, 1978) but these findings have yet to be exploited in an applied context. An inherent problem with this general approach is the sexual selectivity of such lures. Traps scented with male coyote (*Canis latrans*) urine for example catch significantly fewer female coyotes (Gier, 1975). In pest control it can be important to attract females, in particular, in order to limit the reproductive rate but in other applications, such as the cropping of a wild species, the opposite can be true, particularly with polygamous species where females can be spared at the expense of males thereby maintaining a large breeding population. In this way the fishermen of the Mississippi use caged female catfish (*Ictalurus punctatus*) as decoys to catch the males (Timms and Kleerekoper, 1972). A sound knowledge of the ethology of the target species is a prerequisite for any attempt to use chemical signals as attractants for traps since related species can react in opposite ways. The vole (*Microtus townsendii*) and the musk rat (*Ordrata zibethicus*) are both cricetid rodents but whilst the former is attracted to traps by odours of conspecifics (Boonstra and Krebs, 1976), the latter is repelled during the breeding season (van den Berk and Muller-Schwartze, 1984). Musk rats set up breeding territories and therefore avoid other adults; they will, however, visit treated traps in autumn and spring (Ritter *et al.*, 1982).

The prolonged diestrus (Whitten, 1959) and increased frequency of pseudo-pregnancy (van der Lee and Boot, 1955, 1956) shown by female mice housed in large groups without males might have promise as a control method were it not for the practical difficulties inherent in treating an area with female odour, whilst masking the smell of any male attracted to the apparent superabundance of females. Stoddart (1980) has similarly drawn attention to the practical difficulties of using the pregnancy blocking ('Bruce') effect that results from exposure to strange male odour (Parkes and Bruce, 1962) in that even under laboratory conditions only 70% of females succumb and these animals are soon back in breeding condition.

A more promising development for improving rodent control stems from the research of Galef and his co-workers (Galef and Wigmore, 1983; Galef *et al.*, 1984, 1985; Galef and Stein, 1985). They found that rats were far more likely to eat a novel diet following interactions with conspecifics that had consumed the same food. After such interactions the naïve rats also had a reduced tendency to learn an aversion to the novel diet if it was subsequently paired with poison. The smell of the diet on the experienced rat was not sufficient for this effect (Galef *et al.*, 1985; Galef and Stein, 1985); it had to be paired with the odour of carbon disulphide or carbonyl sulphide, which are food breakdown byproducts present in the breath of the rat. Further experiments revealed that simply adding carbon disulphide to bait markedly enhanced its consumption by mice (Bean *et al.*, in press) and attenuated aversion learning to novel diets in rats (Galef *et al.*, 1988). Carbon disulphide mediates social influences on diet selection.

The last major category of potentially useful chemical signals are those that convey alarm. Elephants frequently damage trees whilst feeding, either by knocking them over or by de-barking them. A partial solution is to reduce elephant numbers by culling but this still leaves the problem of protecting specimen trees in specific areas. In the

African elephant (*Loxodonta africana*) the temporal scent glands are particularly active in stressed or aggressive individuals and it has been suggested that this secretion could act as a repellent (Gorman, 1986). When disturbed, black-tailed deer (*Odocoileus hemionus columbianus*) produce a chemical alarm signal from their metatarsal glands (Muller-Schwartze *et al.*, 1984) but, unfortunately, in the field its repellent effects are transitory (Rochell *et al.*, 1974). Rats avoid the odour of dead conspecifics (Carr and Kennedy, 1979) and the urine of frightened rats (Schultz and Tapp, 1973), but again these effects only last for short periods. As chemical signals convey relatively crude temporal information, i.e. the danger may have passed although the signal remains, it is not surprising that alarm odours merely increase vigilance for short periods unless reinforced by auditory or visual signals indicative of imminent danger.

Humans have traditionally mimicked the calls of quarry species to attract their prey, and rabbit hunters or wildfowlers still use this technique today. Numerous ingenious methods have been employed to simulate the correct sounds: for example 'a fairly realistic wigeon whistle can be made by soldering two cartridge heads together and sucking, not blowing' (Marchington, 1977). The calls of female blue grouse (*Dendragapus obscurus*, Stirling and Bendrell, 1966) and quail (*Colinus* sp., Levy *et al.*, 1966) will lure breeding males whilst tape recorded song has been used to attract territorial species such as warblers (*Vermivora* sp., Murray and Gill, 1976). Deer calls have aided the location of deer fawns (Diem, 1954) and lion (*Panthera leo*) sounds have lured lions to a carcase for capture by darting (Smuts *et al.*, 1977). Experiments have also investigated the possibility of using the ultrasonic cries emitted by rodent pups when separated from their mother, as attractants for rats and mice (J. Smith, unpublished data).

Many species of bird emit 'distress calls' (Perrone and Paulson, 1979; Inglis *et al.*, 1982) when seized by a predator, or by a scientist armed with a tape recorder, and the broadcasting of such sounds has for some time provided the basis for bird scaring on airfields (Brough, 1968; Blokpoel, 1976). These calls are now used in other contexts. Starling (*Sturnus vulgaris*) roosts have been shifted from woodland and urban areas by broadcasting starling distress calls over four or five consecutive nights as the birds arrive at the roost (Brough, 1969). Gull roosts on a Scottish reservoir have been dispersed by playing gull distress calls in a similar fashion (Monaghan, 1984). In these situations the birds are looking for a safe place to rest and any indication of danger is liable to shift them. The birds become more tenacious when there is more at stake: 'When distress calls were played by pest control officers through the streets of a town in northern England plagued by nesting gulls, the gulls responded by a vigorous and noisy defence of their nesting sites' (Monaghan, 1984). Similarly, hungry birds are often willing to take risks in order to gather food. Summers (1985) found that while distress calls scared starlings from cherry orchards, the birds nevertheless returned very quickly and were easily able to obtain their daily requirement of food from the orchards. Distress calls, used in conjunction with pyrotechnic devices, have been successfully employed on an experimental basis to keep a refuse tip clear of gulls (J.B.A. Rochard, unpublished report), but the large manpower input required means that this technique is unlikely to be adopted by commercial operators. Clearly we have only just begun to look at the potential uses of acoustic signals. A vast number of, particularly mammalian, sounds remain to be investigated.

There are very few examples of the use of visual intra-specific signals probably because these are usually much harder to replicate than acoustic or chemical signals. Wildfowlers use decoys as a means of attracting their quarry and it has become clear that the postures depicted by decoys can be important. Krebs (1974) found that great blue herons (*Ardea herodias*) preferred to land with models of herons in the upright hunting posture rather than with hunched roosting models. Also Drent and Swierstra (1977), using models of geese either in the head down or head up posture, showed that skeins of barnacle geese (*Branta leucopsis*) choose to land with the model flock containing the greatest number of head down or 'grazing' decoys. In both cases the authors suggest that conspecifics flying overhead estimate the attractiveness of an area for food by the proportion of flock members on the ground that are actually feeding, i.e. hunting or grazing. Preference is lowered by the absence of postures indicative of high food density.

Two other studies have attempted to reduce the attractiveness of an area by the use of artefacts that simulate visual alarm signals. When a goose is alarmed, the body becomes angled upwards, the neck is extended vertically and, just before the bird takes off, there is a rapid side-to-side motion of the head. Inglis and Isaacson (1978) made three-dimensional models of geese in this posture that had heads which vibrated in the wind. Using these models they were able to control to a large degree where skeins of dark-bellied brent geese (*Branta bernicla bernicla*) landed (see also Figure 2.3, Chapter 2). It was subsequently found (I.R. Inglis *et al.*, unpublished data) that simple silhouettes of geese in this posture were as effective providing that the heads could still oscillate in the wind. The second study (Inglis and Isaacson, 1984) involved attempts to simulate a visual alarm signal used by the woodpigeon. This is discussed in a later section.

4.2.2 The use of interspecific signals

Many animals respond to the signals used by other species; these serve as attractants to predators and as repellents to potential prey. Carnivores can undoubtedly learn to associate odour cues with prey (Apfelbach, 1973) and attempts have been made to use a conditioned 'prey odour-illness' association to reduce coyote (*Canis latrans*) predation of domestic stock. Sheep corpses treated with lithium chloride, a strong emetic, are left out for the coyotes. Although promising results were initially reported (Gustavson *et al.*, 1974; Ellins *et al.*, 1977) later studies have been less successful (Bourne and Dorrance, 1982; Horn, 1983; Burns, 1983). The difficulties appear to be fourfold. First, the subjects have often already learnt that the target animal is a 'safe' prey and are therefore less likely to associate it with the illness (Kalat, 1977). Second, the coyote has to generalize from the taste–illness association to an odour–illness association which is learnt only with difficulty (Hankins *et al.*, 1973). Third, the taste-odour–illness association has to inhibit killing, a food-seeking behaviour, and it has been found that the taste aversions affect such behaviours far less than actual food consumption (Garcia *et al.*, 1970). Last, coyote pups do not learn aversions to flesh treated with lithium chloride if it has been regurgitated by their mother (Burns, 1980).

Predators are often attracted more strongly to the odours emitted by frightened or otherwise stressed prey. For example, cats prefer pellets tainted with the odour of 'frightened gerbil' to pellets of 'plain gerbil' flavour (Cocke and Theissen, 1986) and sharks show increased excitement when target fish are harassed by the experimenter (Tester, 1963). There is therefore some ethological justification for the advice 'not to panic' when dealing with potentially harmful animals.

Predator odours are examined by herbivores at close range (Muller-Schwarze, 1974) and there is evidence (Mech, 1977) that they often choose to leave a 'predator area' rather than risk being preyed upon. Sullivan *et al.* (1985a) investigated the effectiveness of a wide range of predator faeces, urine and secretions from anal glands, in suppressing feeding damage by snowshoe hares (*Lepus americanus*). The faeces of lynx (*Lynx canadensis*) and bobcat (*Felix rufus*); the urine of lynx, bobcat, wolf (*Canis lupus*), fox (*Vulpes vulpes*) and wolverine (*Gulo gulo*); and the anal gland secretions of the weasel were found to be the most effective. The lack of any response to domestic dog urine and to 2-methylbutyric acid showed that the hares were not simply responding to a novel odour. Unfortunately the effects were relatively short-lived (up to seven days); but, as the authors point out, the experimental procedure allowed very little control of the concentration of the active ingredients in a given sample and no control over the evaporation rate of the volatile compounds. Another study (Sullivan and Crump, 1964) found that certain compounds from the anal glands of stoat (*Mustela erminea*) and mink (*Mustela vison*) completely suppressed snowshoe hare feeding for at least 38 days; it is suggested that this effect persisted because these compounds polymerized to some degree and hence were less volatile.

Black-tailed deer (*Odocoileus hemionus columbianus*) likewise avoid feeding on plants coated with faeces and urine from a range of predators (Melchiors and Leslie, 1985; Sullivan *et al.*, 1985b). Although all preparations reduced deer damage relative to controls (e.g. human urine), those from the natural predators of the deer (i.e. cougar (*Felix concolor*), coyote and wolf) were, not surprisingly, the most effective. The efficacy of the faecal extract was correlated with the concentration of faeces by weight in the initial formulations until 30% had been reached, after which there was no additional effect. As Sullivan *et al.* (1985b) state 'the active repellent components of predator odours ... may be suitable for encapsulation in controlled-release devices which could provide long-term protection for forest and agricultural crops'.

For obvious reasons predators tend not to make loud acoustic signals whilst hunting. They are however attracted by the sounds made by prey when seized: for example, rabbit distress screams are successfully used to hunt foxes (Morse and Balser, 1961). It is also sometimes possible to manipulate the behaviour of a non-predatory species by broadcasting the alarm signals of others with which it associates. As we have seen, dark-bellied brent geese use a visual rather than an acoustic alarm signal. However, they can be scared from fields by playing the distress call of the curlew (*Numenius arquata*), a wader that frequently feeds in the same areas as the geese (I.R. Inglis, unpublished data).

The use of models of predators to deter birds and other pests is traditional. Man is the major predator of many pest species and as early as 1888 there was available in Britain a clockwork-driven effigy of a man equipped with a gun that automatically fired blank cartridges ('The Millichamp and Sons Patent Automatic Flash Lamp Fox Scarer', Boothroyd, 1988). Modern versions of a 'pop-up' hunter are designed to work in conjunction with propane gas guns that periodically emit a loud noise resembling a shot (Inglis, 1980; Cummings *et al.*, 1986).

Replicas of birds of prey have been used in many situations to deter bird pests. Radio-controlled model aircraft shaped like hawks have been flown on several airfields (Blokpoel, 1976) but most often the model of the raptor is suspended from thin wires or slung beneath helium-filled balloons. The degree of protection afforded to crops by this technique varies widely. For example, Meylan and Murbach (1966) reported negligible damage reduction when trying to protect sunflowers from greenfinches (*Carduelis chloris*), while · Conover (1984) achieved good results on corn fields attacked by red-winged blackbirds (*Agelaius phoeniceus*) and common grackles (*Quiscalus quiscula*). The effectiveness of models of birds of prey can be improved by showing the raptor apparently grasping one of the pest species. Kruuk (1976) reported that lesser black-backed gulls (*Larus fuscus*) and herring gulls (*Larus argentatus*) were more wary of predator models after seeing them beside dead gulls. Conover and Perito (1981), however, found no such enhancement when a dead starling was placed under a model owl, but did obtain a marked improvement when starlings saw the replica with a live starling tethered to its feet. There are obvious ethical implications in this approach (which would not be permitted in Britain) and hence the next logical step was to test the effectiveness of a model raptor grasping an animated model of the pest species. Conover (1985) mounted a plastic model of the great horned owl (*Bubo virginianus*) on top of a model of the American crow (*Corvus brachyrhynchos*). In one version the crow replica had wings that could flap in the wind whilst in another an electric motor flapped the wings for approximately 1 minute in each 15 minutes. Whereas the owl model on its own did not protect vegetables from crows, both of the animated versions significantly reduced damage levels.

The absence of significant predation pressure from a particular species can result in a prey population that fails to recognize the predator when faced with it. The red-backed shrike (*Lanius collurio*) is absent from northern and southern parts of the range of the pied flycatcher and Curio (1978) compared the shrike mobbing behaviour of birds from shrike-free areas with that of birds from parts where shrikes were common. The former group, unlike the latter, showed no or very little mobbing to a model shrike although both groups vigorously mobbed models of the ubiquitous tawny owl (*Strix aluco*). It therefore seems of little use attempting to deter birds using models of raptors not normally found within the area.

Predator recognition in birds can be complex. Falconers are able to tell at a glance whether or not their hawk is 'in the mood' to hunt (said to be 'sharp set'), but are less certain as to the precise visual cues they use to make this judgement. Hamerstrom (1957) compared the tendency of birds to mob a stationary red-tailed hawk (*Buteo jamaicensis*) when it was either well fed or sharp set. Although the hawk's behaviour appeared to be very similar in both cases, the mobbing was far more intense when the raptor was sharp set.

Hamerstrom suggested that the eyes appeared rounder when well fed, and the head feathers tended to be flatter and the wings held slightly higher when sharp set. These experiments clearly demonstrate the high degree of complexity of which the predator recognition process is capable. It is therefore somewhat strange that birds can be repelled by stimuli that only crudely resemble a small part of the predator, namely its eyes.

The wings of many insects bear circular patterns or 'eyespots' that resemble the vertebrate eye and are used in displays which frequently elicit flight in the passerine predators of these creatures. Blest (1957) showed that butterflies (*Aglais urticae*) which had the patterns removed from their wings were unable subsequently to defend themselves from yellow buntings (*Emberiza citrinella*) and reed buntings (*Emberiza schoeniculus*). Kerlinger and Lehrer (1982) similarly found the presence of eyes to be important in controlling the mobbing of owl models by sharp-shinned hawks (*Accipter striatus*). Replicas with eyes were mobbed more intensely and at a greater distance than were those lacking eyes. Eye-like patterns can trigger avoidance in a range of vertebrates (Blest, 1957; Scaife, 1976; Coss, 1978; Jones, 1980) and Inglis *et al.* (1983) looked at the possibility of using such patterns to deter hungry starlings from food. They found that the presence of a pair of 'eyes' (each eye being a 2cm diameter circle containing a 1cm diameter solid pupil) drawn in black on a white background caused an 88% reduction in the average time starlings spent on a food trough and a 65% decrease in actual feeding time. The size, shape, number, colour and other parameters of the patterns were varied and the most effective combination was as aversive as a model of an owl (see Figure 4.3). Unfortunately there was also evidence that the responsiveness of the birds declined over the five days of the experiment. It is claimed that eye patterns are used with some success in a variety of situations. For example, huge 'eyes' are now painted on the engine nacelles of jets belonging to the All Nippon Airways in order to reduce bird strikes (S. Shima, unpublished data) whilst small 'eyes' have been sewn onto jogger's caps to deter nesting magpies (*Gymnorhina* sp.) from attacking keep-fit enthusiasts (Eisner, 1985).

An important area of interspecific communication concerns the relationship between us and our companion animals (Anderson, 1975; Fogel, 1981; Katcher and Beck, 1983 and see Chapter 8). There is often a high degree of empathy between owner and pet. Eighty-one per cent of respondents in a survey stated that their pets 'tuned in' to the feelings of members of the family (Cain, 1983). The animals were reported to have experienced diarrhoea, stomach upsets and epileptic attacks at times when family relationships were strained. Similarly 77%

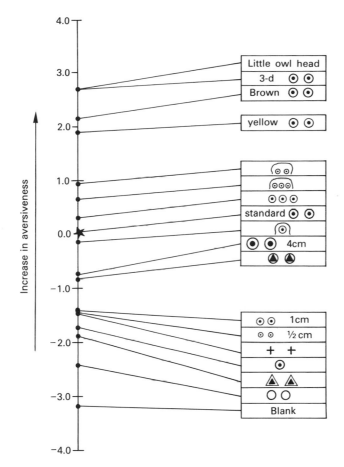

Figure 4.3 The degree of aversiveness of simple 'eye-like' patterns to starlings (modified from Inglis *et al.*, 1983). The patterns are ranked in comparison with the standard stimulus which by definition has the value zero. All 'eyespots' were 2cm in diameter with a 1cm diameter black 'pupil' unless otherwise stated. Stimuli are significantly different at the 5% level (one-tailed) when they differ by more than 1.6. Eyespots with coloured 'irises', and hemispherical plastic eyes proved to be as aversive as a head of a Little owl (*Athene noctua*).

of those whose pet had died described the loss as a period of sadness, grieving and mourning. Too often, however, a failure to understand the signals given by our pets can lead to problems. For example, the domesticated dog is still a predator and dog bites are a major problem affecting millions of people each year (Borchelt *et al.*, 1983). It is generally believed that dogs were domesticated from the grey wolf

(*Canis lupis*) (Messent and Serpell, 1981) and 'In many ways, dogs relate to their owners as dominant members of the pack; many of the same postures and vocalizations used by wolves are seen to take place between people and their dog. The "play-bow", crouch and whimper are all commonly understood by people as if the subordinate "wolf" were communicating with the alpha animal' (Katcher and Beck, 1983; p178). Problems arise when, because of lack of discipline, this dominance relationship is unclear or the dog is allowed to become the alpha animal. As Fox (1975) points out, the effective way to instil discipline is to use the same channels of communication as the dog: 'a direct stare is threatening; a rough voice, a growling command; seizure and closure of the muzzle, analogous to muzzle-biting; seizing the scruff of the neck and shaking and pinning the animal down to the ground by its muzzle' (Fox, 1975; p. 45). This type of direct communication, coupled with a regime of rewarding submissive behaviour and avoiding opportunities for displays of dominance has been refined into a treatment plan for dogs that attack their owners (Voith, 1981). A more detailed discussion of behavioural problems in companion animals like dogs can be found in Chapter 8.

When herding ungulates we customarily use manoeuvres which closely resemble those employed by dominant animals to keep herds together or to separate a female from the herd (Walther, 1984). The bullfighter alternates between adopting a large profile (with the help of his cloak) towards the bull which imitates the threat posture of bovines, and a sideways smaller profile that is far less aggressive to calm the animal. Presumably by trial and error people learnt that these particular movements were most effective in controlling their herds.

The nature of the communication between shepherds and their dogs would appear at first sight to be outside the scope of this chapter since the animals are taught to respond to completely artificial signals created by the shepherds. However, studies of the acoustic signals employed (Vines, 1981; McConnell and Baylis, 1985) reveal that this is not the whole story. As expected, the sounds used for the commands 'go left' and 'go right' differ markedly between shepherds, but the signals used for the other two basic commands, 'fetch' (i.e. push the flock towards me) and 'stop', are very stereotyped. The 'fetch' command is a series of short repeated notes, often increasing in frequency at the end of each note; whilst the 'stop' command is a single, long note ending on a descending frequency. It is unlikely that any similarity has resulted simply from dog training methods being passed from shepherd to shepherd since 'if cultural transmission alone acted ... we would expect each of the four basic working whistles to show equivalent stereotypy' (McConnell and Baylis, 1985). Furthermore, verbal stop and fetch signals varied phonetically but showed the same duration and note structure as the related whistles.

Unlike the directional signals, the fetch and stop commands trigger greater and less activity respectively, and McConnell and Baylis (1985) note that similar acoustic structures are used to speed up and slow down horses. Short pulsed notes are common both in canids (Fox and Cohen, 1977) and primates (Moody and Menzel, 1976), and result in increased activity in the receiver. Conversely, continuous descending notes are often used to depress activity. McConnell and Baylis (1985) argue that the correlation of signals 'is a function of both producer and receiver; that primates, including human beings, tend to produce pulsed calls in stimulating contexts, and that the same acoustic structure increases the chance of higher activity in the canid receiver'.

Other workers (Collias, 1960; Morton, 1977) have pointed to similar structural stereotypy in sounds used in 'hostile' and 'friendly' contexts. 'Stated simply birds and mammals use harsh relatively low frequency sounds when hostile and higher frequency, more pure tone-like sounds when frightened, appeasing, or approaching in a friendly manner' (Morton, 1977). Why this is so is not clear. Morton (1977) suggests that low frequency sounds are used in hostile situations because, as we have mentioned, there is a direct relationship between large size and the production of low sound frequencies. Conversely high sounds tend to be produced by small, and particularly young, animals hence it may be that such sounds reduce aggression by eliciting 'parental-care type responses'. Whatever the reason, the existence of a fundamental relationship between acoustic signal structure and signal meaning suggests that there might be similar structural rules for signals in other modalities, and that it may be possible to synthesize novel signals which are more effective in eliciting the appropriate responses than the normal ones; a possibility to which we shall return.

4.2.3 Habituation and extinction

The major problem in the practical use of signals is a gradual decline in their effectiveness with repeated use, as the target animals become less responsive to them. There can be difficulty in accurately mimicking natural signals and therefore the waning in responsiveness could result from deficiencies in the artificial signal. However, this is not the whole story since similar declines can occur in response to highly salient natural signals. For example, the mobbing responses of chaffinches (*Fringilla coelebs*) decrease over time even when elicited by a live owl (Hinde, 1954), and the repellent effect of conspecific blood (Stevens and Gerzoy-Thomas, 1977) wanes sufficiently to allow cannibalism to take place in rats (Carr and Kennedy, 1979).

There are several possible explanations for a decline in responsiveness. The animal may cease to recognize the stimulus as a result of sensory adaptation. However, the subjects often respond to the signals in a different manner thereby indicating that they do still perceive them. The lack of response might simply be a result of muscle fatigue and yet the same muscles are used perfectly normally in other movements. The decline in effectiveness therefore appears to be the result of changes in the central nervous system. The animal learns not to respond to stimuli which are without biological importance (e.g. do not signal an actual attack by a predator). Traditionally, such learning has been split into two processes according to how the original response developed. If the initial response to the stimulus was 'spontaneous' then the decline is said to result from habituation. If, however, the animal had been trained to respond to the stimulus by pairing it with reward or punishment, then the decline that follows the withdrawal of such reinforcement is called extinction. The two processes have many features in common (Kling and Stevenson, 1970) and the division appears somewhat arbitrary since seemingly spontaneous responses may have previously been learnt by the animal even if not as a result of experimental training in the laboratory.

Thompson and Spencer (1966) listed nine characteristics of habituation. Both the strength of the response, and the probability that it will be elicited, decline with the number of stimulus presentations. The more frequent the presentations, the more rapid is the habituation. If the stimulus is withheld for a period and then restarted, the response strength increases. This effect is called 'spontaneous recovery' and is greater the longer the period without the stimulus. For example, Hinde (1954, 1961) conducted a series of experiments investigating the habituation of the mobbing response of the chaffinch. When first confronted with a stuffed owl the finches flew around it making a 'chink' call. The rate of calling fell rapidly over a 30 minute period but then recovered after the owl had been withdrawn for two hours. In general, the longer the period without the owl the greater was the degree of spontaneous recovery, although even after a 24 hour gap the chaffinches only emitted about half the number of calls recorded on the first trial. The results of such studies have obvious implications for the way in which we should use signals. Unfortunately there is still a widespread belief amongst farmers and others that the more frequent a signal like a distress call is used, the greater will be its effect, even though advisory leaflets (Inglis, 1987) stress that this is not the case.

Many of the other characteristics listed by Thompson and Spencer (1966) have practical as well as theoretical importance. For instance, they explain that continued exposure to the signal even after the response has ceased will reduce any subsequent spontaneous recovery of that response. Continuing to use devices after the animals have stopped responding to them can have other deleterious effects. For example, once birds have become thoroughly habituated to certain visual bird scarers they may use them as indicators of good food supplies (I.R. Inglis and A.J. Isaacson, unpublished data quoted in Feare *et al.*, (1990), a meaning very different from the original message!

Thompson and Spencer (1966) also state that animals which have habituated to a particular stimulus will exhibit weaker responses to other stimuli that resemble it. Conversely, the response may show partial recovery if a new stimulus is presented. The latter effect is strongest if the new stimulus is very different from the previous one but it is possible to get some improvement by increasing the intensity and/or changing the location of the existing signal. For example, Shalter (1975, 1978) found that a slight alteration in the position of a model predator revived the responsiveness of captive jungle fowl (*Gallus gallus spadiceus*) and wild pied flycatchers (*Ficedula hypoleuca*) after habituation to the model in the previous position.

The speed at which extinction occurs to a learned response appears to depend on the degree of difference between the conditions of the learning and extinction stages (Mackintosh, 1974; Chapter 8). An animal which has learnt to press a bar for a food reward that is delivered each time the bar is pressed, rapidly ceases to respond once the food is withdrawn. However, if during the learning phase the bar had to be pressed 20 times for each reward then bar pressing persists for much longer after the food has ceased. In the second situation the animal has learnt that bar presses often result in no reward and it takes far longer to be sure that the circumstances have changed than it does in the first situation where the animal has learnt to expect a reward every time. An animal will therefore cease responding most rapidly if the signal is usually reliably, and quickly, followed by an event which is absent in the applied context. For example, the mating stance of oestrous pigs can be elicited using aerosol preparation of boar odour (a mixture of androstene steroids; Booth, 1980) but this signal must be rapidly followed by others (e.g. pressure on the back, audible grunts) that are normally associated with the close proximity of a boar (Signoret, 1970). Signals that are only occasionally paired with the relevant event are effective for longer. Thus you would not expect always to see the sender each time you encountered a territorial scent mark, and indeed the responses of Ring-tailed lemurs (*Lemur catta*) to the odour of a foreign troop wane slowly and are reinstated if that odour source is temporarily removed and then replaced (Mertl, 1977).

The signals that lend themselves most readily to practical use are those that herald some dangerous, or otherwise stressful, event. Animals avoid the event by fleeing as soon as the warning signal is received. There is a vast literature concerned with this phenomenon of avoidance learning (Mackintosh, 1974) and one fact to emerge is that the warning stimulus can retain potency for very long periods in spite of the fact that, once the warning has been recognized as such, the animal never receives the aversive stimulation to reinforce the avoidance response. It has been argued (Hilgard and Marquis, 1940) that the animal forms an expectancy of the unpleasant event contingent on the warning stimulus and therefore it is the *omission* of the expected event which reinforces fleeing. This expectancy will not be weakened unless the animal decides to ignore the signal and discovers that the predicted event does not in fact occur. Yet individuals that ignore warnings tend not to pass on the genes for such behaviour. Once the warning signal has been recognized it matters little whether the aversive reinforcement is present or absent for, as far as the animal is concerned, there is no difference between the learning and any possible extinction conditions, therefore the avoidance persists.

Interspecific signals like predator odours should be effective over relatively long periods since to come into contact with the source of the odour could be fatal. Mech (1977) has shown how deer use wolf scent marks to enable them to avoid wolf territories. Indeed Kruuk and Sands (1972) have suggested that some predators use common latrines, or middens, to localize, and ensure the rapid disposal of, faeces which might otherwise alert their prey. The applied use of predator odours can result in long-term repellency. As we have seen, lodgepole pine seedlings were completely protected from damage by snowshoe hares for at least 38 days by treatment with a component of mustelid scent gland secretion (Sullivan and Crump, 1984). However, repellency is often less than absolute. For example, traps treated with weasel scent do still catch small rodents albeit less frequently than do untreated traps (Stoddart, 1976).

Intraspecific warning signals should also retain effectiveness for long periods and indeed the response of nestling great tits (*Parus major*) to the adults' alarm call wanes comparatively slowly (Ryden, 1978). However, not all aversive signals result in conspecifics avoiding the source of the signal. The 'distress scream' is aversive in that chaffinches (Thompson, 1969) and jackdaws (*Corvus monedula*)·

(Morgan and Howse, 1973) will hop on to a perch or peck a key in order to avoid hearing such calls. Nevertheless, in the wild, individuals of these species approach the source of this aversive stimulation and investigate its cause. If a bird flees in response to an alarm call it obviously has little opportunity to discover whether it has been deceived or not. Mobbing behaviour will, however, provide information on the source of the signal so that we might expect rapid habituation of response to distress calls. Yet this appears not to be the case; distress calls are as effective as alarm calls (Inglis *et al.*, in preparation). It may be that the act of mobbing actually enhances fear. One explanation of mobbing behaviour is that a bird learns to fear an object once it has witnessed conspecifics mobbing it (Curio, 1978). There is some experimental evidence for this view. Curio *et al.* (1978) have shown that 'observer' blackbirds (*Turdus merula*) mobbed a non-raptorial bird more strongly as a sequence of witnessing another blackbird mob strongly at the site of presentation (this bird was, in fact, mobbing a model of a predator hidden from the observers who could see only the non-raptorial bird). The strength of the enhanced response was comparable to that elicited by a genuine predator. Further, Curio *et al.* (1978), by using the same techniques, tutored blackbirds to mob a novel stimulus, a plastic bottle, although the response was somewhat weaker than normal. This is not a curious effect that occurs only under experimental conditions. On British airfields bird scaring is usually achieved by playing distress calls from loudspeakers mounted on a vehicle. After some time the birds start to respond as soon as the vehicle appears and before the calls are played. This reaction is specific to the vehicle carrying the loudspeakers (N. Horton, personal communication); it has become the 'predator'.

A knowledge of learning theory can therefore help us to increase the effectiveness of signals used in an applied context, first, by providing guidelines on the sorts of signals that are likely to be effective, and second, by suggesting operating procedures for the optimal use of those signals. We now turn to another way to improve performance, that is to modify the signal itself.

4.2.4 Signal modification

Animals do not usually respond to the entire stimulus pattern that accompanies some event of biological importance. Rather, as we have seen, certain aspects are far more important than others in triggering the appropriate behavioural response. Such cues may elicit the normal behavioural response even when presented in the absence of the other stimuli that usually accompany them; these cues have been called 'sign stimuli' (Russell, 1943). A now classic series of experiments (Tinbergen and Perdeck, 1950; Tinbergen, 1953) using models of the heads of herring gulls (*Larus argentatus*) revealed that the characteristics important in eliciting the begging responses of chicks were a long, thin object, pointing downwards, with a red patch near the tip. A red knitting needle with white bands near the tip was found to be more effective in eliciting begging responses than was a model of the normal gull's head. The needle formed a 'super-normal stimulus' by combining in an exaggerated form the important visual cues that are both necessary and sufficient to elicit begging. Presumably gulls do not possess super-effective knitting-needle bills since the bill has to fulfil other functions which may not efficiently be served by such a shape (e.g. feeding). The normal bill is adequate in that gull chicks do not need to peck more intensely. Super-normal responses to modified signals have now been found in a number of species (e.g. butterflies, Magnus, 1958; humans, Gardner and Wallach, 1965) and this has fuelled speculation that super-normal stimuli could prove useful in an applied context (Brémond, 1980; Inglis, 1980). However, there has been little research directed at this possibility and so far there are only a few examples of super-normal stimuli that could be of use to us.

The woodpigeon is an economically important pest feeding extensively on a wide range of arable crops. Shooting is the common control method and experiments were conducted (Murton *et al.* 1974) to investigate the optimal use of decoys (i.e. preserved woodpigeon bodies). In the course of this work it was discovered that of those birds which circled the decoys, 54% landed near closed-winged decoys and only 4% close to open-winged decoys. This might mean that open-winged decoys are the less attractive or, as Murton (1974) speculated, that open-winged decoys are in fact repellent, because they present some intraspecific sign stimuli eliciting flight not shown by the close-winged decoys. The obvious candidates for such sign stimuli are the white wing marks, clearly visible in open-winged bodies but not in closed-winged corpses. The woodpigeon has no alarm call and therefore the wing marks might form a visual alarm signal. Inglis and Isaacson (1984) investigated this possibility. By presenting woodpigeons coming to a preferred feeding area with a series of binary choices between different decoy types, it was confirmed that the wing markings do indeed function as an aversive signal. They then tried to make a super-normal stimulus by modifying the wing marks and discovered that doubling the area of each mark resulted in decoys that were significantly more aversive than the normal open-winged

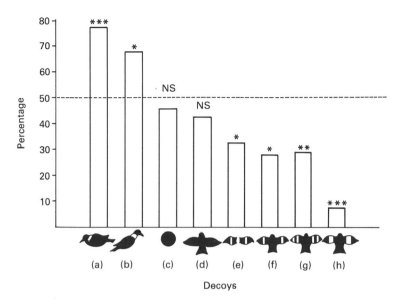

Figure 4.4 The mean percentage of woodpigeons settling with the decoys illustrated as opposed to landing within an area containing no decoys (Inglis and Isaacson, 1984). The asterisks indicate the degree of significance of this preference with respect to the no-preference 50% level (* = $P<0.01$, ** = $P =$ 0.001, *** = $P < 0.001$). The decoys are (a) normal wings-closed, (b) extended-neck wings-closed, (c) a plastic disc of approximately the same area as an open-winged decoy, (d) wings-open with the wing marks painted out, (e) a pair of wings, (f) wings-open with normal wing marks, (g) wings-open with two wing marks, (h) wings-open with enlarged wing marks.

bodies. Figure 4.4. shows the relative degrees of aversiveness shown by the various decoys tested.

Subsequent field trials (Inglis and Isaacson, 1987) demonstrated that open-winged bodies significantly reduced woodpigeon damage to clover over a nine-week period. However, the super-normal decoys were no more effective than the normal open-winged bodies after the initial few weeks of the experiment. As all previous work on super-normal stimuli had involved only short exposure periods it is impossible to say yet whether this finding is of general validity. Why the wing marks, as shown in the static open-winged posture, should be effective is something of a mystery, as normally these white marks are visible only as an oscillating pattern of white bars. It appears that the speed of this oscillation may convey the alarm signal as the wingbeat frequently is significantly greater when a woodpigeon is flushed than when it takes off spontaneously (Inglis and Isaacson, 1984). Devices are being made to simulate this visual stimulus and it will be interesting to compare their effectiveness with that of the open-winged bodies.

There are good reasons to believe that super-normal acoustic signals can similarly be produced (Brémond, 1980) and, with the advent of computerized sound synthesizers and analysers, it is easier to work with acoustic signals than visual ones. The first step is to present modified calls that lack certain characteristics of the normal signal and compare the level of responsiveness to such calls with that usually given to the normal signal. In this way the most important parameters of the signal can be identified. For example, Morgan and Howse (1974) conditioned jackdaws to peck a key in order to avoid hearing a jackdaw distress scream and then presented the birds with recordings that lacked various bits of the normal call. They found that removal of the starting and finishing transients made the stimulus far less aversive. Aubin (1987) analysed the starling distress scream and then measured the responses of flocks to various synthesized calls containing only parts of the normal signal. In this way it was discovered that the carrier frequency plus only the primary modulation was sufficient and necessary to elicit the normal reactions.

Once the crucial features of the signal have been identified the next step is to synthesize a call that emphasizes them. Unfortunately there have been few such experiments. The assembly call of the common American crow (*Corvus brachryrhynchos*) was studied by Richard and Thompson (1978). This call consists of a series of 'caw' sounds which are low, harsh and variable in pitch and timing. The crucial factor was found to be the temporal organization of these 'caws'. By taking other very harsh sounds that normally elicit mobbing by crows and organizing them into short sequences of increasing emission rate, Richard and Thompson (1978) were able to manufacture a call which was 'two to four times more effective than the natural assembly call'. Aubin (1978, unpublished Ph.D thesis quoted in Brémond, 1980) conducted a similar investigation of the territorial song of the skylark (*Alauda arvensis*). After discovering the part of the song that was most effective in producing aggression in the larks, he built a synthetic song, using an electronic synthesizer, that repeated the same motif with a uniform and regular tempo. The resulting song was much simpler than the natural song and super-normal in eliciting threat and fighting displays from larks in the field. This approach holds great promise and Brémond and his co-workers are now trying to synthesize calls to repel starlings, gulls and members of the crow family (J-C. Brémond, personal communication).

It may be possible to obtain an enhanced response not by creating a novel super-normal sound but by simply enhancing one or more parameters of the normal signal above the usual range. For example, the orange-chinned parakeet (*Brotogeris jugularis*) emits a harsh sound whose loudness appears to be correlated with the level of anxiety aroused when a conspecific approaches the nest (Power, 1966). Would a recording of this call broadcast at intensities beyond the capability of the bird elicit an enhanced response from the intruder? The rapidity of calling is also often proportional to the intensity of motivation associated with the signal. The greater the danger, the faster is the alarm call emitted by the Pygmy owl (*Gaucidium passerinium*) (Konig, 1968), and perhaps a signal broadcast with a call rate beyond the capacity of the owl would similarly produce a greater response in conspecifics. The intensity of response often varies with the distance from the source of the signal, and there is evidence that animals use the relative attenuation of frequencies within the signal to judge the distance from the source (Morton, 1986). Extra energy in those higher frequencies that are attenuated the most should therefore deceive the receiver into judging the source to be closer than it is, and to act accordingly. Brémond (1980) has shown that broadcasting a territorial song of the robin (*Erithacus rubecula*), which has been modified in this way, does result in unusually strong responses from conspecifics. Clearly we are only just beginning to investigate the potential of modified acoustic signals for use of pest control and wildlife management.

There is no theoretical reason why it should not be possible to create super-normal chemical signals but we know of no good examples. The strong response of cats to the smell of the essence of 'catnip' (*Nepeta* sp.) is identical to the rubbing and rolling of a female cat in oestrus (Todd, 1962). The active ingredient appears to be *cis trans*-Nepetalactone which is effective at only 1 part in 10^9 to 10^{11} (Eisenbraun *et al.* 1980). A compound with a similar structure is found in the Chinese gooseberry (*Actinidia polygoma*) and this elicits the same responses in cats (Sakai *et al.* 1980). It has been suggested (Todd, 1963) that these compounds may resemble as yet undetected semiochemicals used by the cat.

There is an enormous variety of chemical signals and the basic problem when working with animals other than humans it the impossibility of obtaining adequate descriptions of aroma potency and quality. As MacLeod (1980) states 'it is certainly true that a cat, for example, can readily express like or dislike concerning a particular odour but it is impossible to gain any data from such mammals concerning detailed odour qualities'. The vast majority of our information therefore relates to humans. An additional problem is that there are no totally convincing theories to explain the relationship between chemical structure and our perceptions of the odour. Compounds with similar structures can have similar smells but equally they can often be different, and compounds with completely different structures can have the same smell. These difficulties are illustrated by some of the results of research carried out on the relationship between chemical structure and the odour of musk. Compound A (see below) smells strongly of musk and compound B, which has a very similar structure, has a weak musk odour (Theimer and Davies, 1967). However compound C has a very different structure from A and B and yet also smells strongly of musk, whilst compound D, having a structure very like C, emits a powerful aroma of urine (Prelog *et al.* 1945), a crucial difference when trying to concoct a perfume!

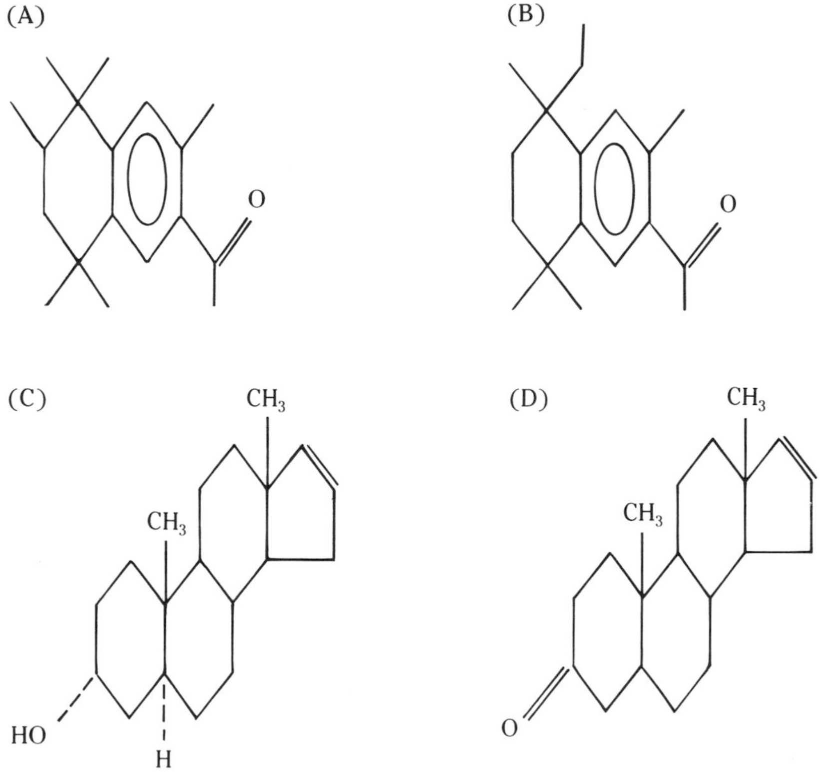

Figure 4.5 Relationship between the chemical structure and the odour of musk.

There appears to be little likelihood at present of being able to synthesize novel super-normal chemical signals. However, it is often possible to achieve useful results by increasing the concentration of the chemical well above the normal level. For example, American construction firms normally check for leaks in new pipelines with lasers and ultrasonic equipment but a far more cost-effective method is to use the responses shown by turkey vultures (*Cathartes aura*) to chemicals emitted from corpses. If a high concentration of ethyl mercaptan is pumped into the unchecked pipeline some will escape and be detected by the vultures who circle and then descend to remain in close proximity to the leak. Therefore to detect a faulty weld the engineer simply requires a pair of binoculars and a portable anemometer (Stager, 1964).

4.3 CONCLUSION

We hope this review has demonstrated that, although we may never achieve the fluency of Dr Doolittle (Lofting, 1922), unravelling the meanings of animal signals opens up great possibilities for controlling the behaviour of animals. Too often success has been limited because we have tried to understand the use of particular signals from a human standpoint rather than from the results of detailed ethological analyses. There is a need to guard against anthropomorphic biases.

It must be recognized, of course, that in economic terms the examples given in this chapter pale into insignificance when compared with the effort expended on trying to manipulate other people. Huge industries are based on the desire to influence the behaviour of others by changing the way we look and smell. As we have developed a uniquely sophisticated system of auditory signals involving manipulation via tone, rhetoric, oratory, etc., it is perhaps not surprising that until relatively recently little scientific attention was paid to our visual and chemical signals.

It now appears that we employ characteristic, expressive gestures which develop with minimal cultural interference (Ekman and Friesen, 1969, 1975; Eibl-Eiblesfeldt, 1972), and a superficial survey of 'body language' has become an obligatory part of most management courses. Cosmetics have traditionally been used to hide the visual signs of ageing and to accentuate those facial features used in communication. It has been suggested that the eyes are painted to increase apparent size and thereby mimic 'baby faces' which are attractive to both sexes after puberty (Fullard and Reiling, 1976), whilst the lips may be coloured so as to resemble sexually receptive female genitalia (Morris, 1977).

Although we have grown accustomed to regarding ourselves as relatively anosmic, humans can display surprising olfactory acuity. For example, adults can discriminate the odour of their sexual partners (Hold and Schleidt, 1977) and mothers recognize the smell of their newborn infants (Porter and Moore, 1981). Most of the volatiles used in perfume are employed as olfactory signals by other species: indeed it has been claimed (Wood, 1978) that no successful fragrance has ever been compounded without the use of a musk. The cultural correlation of strong odours with dirtiness and unattractiveness has overridden their more primitive association with fear, sex and dominance.

The attempted manipulation of our own kind by the use of non-verbal signals could have far more sinister implications than the use of the signals of other species in animal husbandry or wildlife management.

REFERENCES

Anderson, R.S. (ed.) (1975) *Pet Animals and Society*, Baillière Tindall, London.

Andrew, R.J. (1961) The displays given by passerines in courtship and reproductive fighting: a review. *Ibis,* **103**, 315–48.

Andrew, R.J. (1963) The origin and evolution of the calls and facial expressions of primates. *Behaviour,* **20**, 1–109.

Apfelbach, R. (1973) Olfactory sign stimulus for prey selection in polecats (*Putorius putorius* L.). *Z. Tierpsychol.,* **33**, 270–3.

Atz, J.W. (1970) The application of homology in behaviour, in *Development and Evolution of Behaviour*, (ed. T.C. Schneirla), pp. 53–74, Freeman, San Francisco.

Aubin, T. (1987) Respective parts of the carrier and of the frequency modulation in the semantics of distress calls. *Behaviour,* **100**, 123–33.

Baldwin, J.T. (1968) The social behaviour of adult squirrel monkeys (*Saimiri sciureus*) in a semi-natural environment. *Folia primat.,* **9**, 281–314.

Baraends, G.P. (1958) Comparative methods and the concept of homology in the study of behaviour. *Arch. Neerl. Zool.,* **13**, 401–17.

Barlow, G.W. (1977) Modal action patterns, in *How Animals Communicate* (ed. T.A. Sebeok), pp. 98–134, Indiana University Press, Bloomington.

Bastock, M., Morris, D. and Moynihan, M. (1953) Some comments on conflict and thwarting in animals. *Behaviour,* **53**, 66–83.

Bateman, J.A. (1973) *Animal Traps and Trapping,* David and Charles, Newton Abbot.

Bean, N.J., Galef, B.J. and Mason, R.J. (in press) At biologically significant concentrations, carbon disulfide both attracts mice and increases their consumption of bait. *J. Wildl. Manage.*

Berk, J. van den and Muller-Schwartze, D. (1984) Responses of wild muskrats (*Ondatra zibethicus*) to scented traps. *J. Chem. Ecol.,* **10**, 1411–15.

Blest, A.D. (1957) The function of eyespot patterns in the Lepidoptera. *Behaviour,* **11**, 209–55.

Blest, A.D. (1961) The concept of ritualisation, in *Current Problems in Animal Behaviour* (eds W.H. Thorpe and O.L. Zangwill), pp. 102–24, Cambridge Univ. Press, Cambridge.

Blokpoel, E. (1976) *Bird Hazards to Aircraft,* Irwin and Co. Ltd., Ontario.

Boeckh, J., Khaisshing, K.-E. and Schneider, D. (1965) Insect olfactory receptors. *Cold Spring Harbor Symp. Quant. Biol.,* **30**, 263–80.

Boonstra, R. and Krebs, C.J. (1976) The effect of odour on trap response in *Microtus townsendii. J. Zool.,* **180**, 467–76.

Booth, W.D. (1980) Endocrine and exocrine factors in the reproductive behaviour of the pig. *Symp. Zool. Soc. Lond.,* **45**, 289–311.

Boothroyd, G. (1988) The clockwork fox scarer. *Shooting Times and Country Magazine,* May 12–18, p. 35.

Borchelt, P.L., Lockwood, R., Beck, A.M. and Voith, V.L. (1983) Dog attack involving predation on humans, in *New Perspectives on Our Lives with Companion Animals* (eds A.H. Katcher and A.M. Beck), pp. 219–231, University of Pennsylvania Press, Philadelphia.

Bouisson, M.F. (1968) Effet de l'ablation des bulbes olfactifs sur la reconnaissance du jeune par sa mère chez les ovins. *Rev. Comp. Anim.,* **2**, 77–83.

Bourne, J. and Dorrance, M.J. (1982) A field test of lithium chloride aversion to reduce coyote predation on domestic sheep. *J. Wildl. Manage.,* **46**, 235–9.

Brémond, J. -C. (1980) Prospects for making acoustic super-stimuli, in *Bird Problems in Agriculture* (eds E.N. Wright, I.R. Inglis and C.J. Feare), pp. 115–20, British Crop Protection Council, Croydon, London.

Bronson, F.H. (1976) Urine marking in mice: causes and effects, in *Mammalian Olfaction, Reproductive Processes, and Behaviour* (ed. R.L. Doty), pp. 119–43, Academic Press, New York.

Brooks, P.H. and Cole, D.J.A. (1970) The effect of the presence of a boar on the attainment of puberty in gilts. *J. Reprod. Fert.,* **23**, 435–40.

Brough, T. (1968) Recent developments in bird scaring at airfields, in *The Problem of Birds as Pests* (eds R.K. Murton and E.N. Wright), pp. 29–38, Academic Press, London.

Brough, T. (1969) The dispersal of starlings from woodland roosts and the use of bio-acoustics. *J. Appl. Ecol.*, **6**, 403–10.

Brown, C.H. (1982) Ventriloquial and locatable vocalizations in birds. *Z. Tierpsychol.*, **59**, 338–50.

Brown, K. (1979) Chemical communication between animals, in *Chemical Influences on Behaviour* (eds K. Brown and S.J. Cooper), pp. 599–649, Academic Press, London.

Burns, R.J. (1980) Effect of lithium chloride in coyote pup diet. *Behav. Neur. Biol.*, **30**, 350–6.

Burns, R.J. (1983) Microencapsulated lithium chloride bait aversion did not stop coyote predation on sheep. *J. Wildl. Manage.*, **47**, 1010–17.

Cain, A.O. (1983) A study of pets in the family system, in *New Perspectives on Our Lives with Companion Animals* (eds A.H. Katcher and A.M. Beck), pp. 72–81, University of Pennsylvania Press, Philadelphia.

Carr, W.J. and Kennedy, D.F. (1979) A response-competition model designed to account for the aversion to feed on conspecific flesh, in *Advances in the Study of Behaviour*, **15**, (eds. J.S. Rosenblatt, C. Beer, M.-C. Busnel and P.J.B. Slater), pp. 245–74, Academic Press, Orlando.

Caryl, P.G. (1979) Communication by agonistic displays: What can games theory contribute to ethology? *Behaviour*, **68**, 136–69.

Caryl, P.G. (1982) Animal signals: a reply to Hinde. *Anim. Behav.*, **30**, 240–4.

Cheney, D.L. and Seyfarth,. R.M. (1988) Assessment of meaning and the detection of unreliable signals by vervet monkeys. *Anim. Behav.*, **36**, 477–86.

Cherry, C. (1957) *On Human Communication*, M.I.T. Press, Cambridge, Mass.

Clutton-Brock, T.H. and Albon, S.D. (1979) The roaring of red deer and the evolution of honest advertisement. *Behaviour.*, **69**, 143–69.

Cocke, R. and Thiessen, Del. D. (1986) Chemocommunication among prey and predator species. *Anim. Learn. Behav.*, **14**, 90–2.

Collias, N.E. (1960) An ecological and functional classification of animal sounds, in *Animal Sounds and Communication* (eds W.E. Lanyon and W.N. Tavolga), pp. 368–91, Amer. Inst. Biol. Sci. Pub. No. 7, Washington, DC.

Conover, M.R. (1984) Comparative effectiveness of Avitrol, exploders and hawk-kites in relation to blackbird damage to corn. *J. Wildl.*

Manage., **48**, 109–16.

Conover, M.R. (1985) Protecting vegetables from crows using an animated crow-killing owl model. *J. Wildl. Manage.*, **49**, 643–5.

Conover, M.R. and Perito, J.J. (1981) Response of starlings to distress calls and predator models holding conspecific prey. *Z. Tierpsychol.*, **57**, 163–72.

Coss, R.G. (1978) Perceptual determinants of gaze aversion by the lesser mouse lemur (*Microcebus murinus*), the role of two facing eyes. *Behaviour*, **64**, 248–70.

Cullen, J.M. (1966) Reduction in ambiguity through ritualization. *Phil. Trans. R. Soc. Lond. B.*, **251**, 363–74.

Cullen, J.M. (1972) Some principles of animal communication, in *Nonverbal Communication* (ed. R.A. Hinde), pp. 101–22, Cambridge University Press, Cambridge.

Cummings, J.L., Knittle, C.E. and Guarino, J.L. (1986) Evaluating a pop-up scarecrow coupled with a propane exploder for reducing blackbird damage to ripening sunflower. *Proc. 12th Verte. Pest. Conf.*, pp. 286–91, University of California Press, California.

Curio, E. (1978) The adaptive significance of avian mobbing. I. Teleonomic hypotheses and predictions. *Z. Tierpsychol.*, **48**, 175–83.

Curio, E., Ernst, V. and Vieth, W. (1978) The adaptive significance of avian mobbing. II. Cultural transmission of enemy recognition in blackbirds: effectiveness and some constraints. *Z. Tierpsychol.*, **48**, 184–202.

Daanje, A. (1950) On locomotory movements in birds and the intention movements derived from them. *Behaviour*, **3**, 48–98.

Dane, B., Walcott, C. and Drury, W.H. (1959) The form and display actions of Goldeneye (*Bucephala clangula*). *Behaviour*, **14**, 265–81.

Darwin, C. (1872) *The Expression of the Emotions in Man and Animals*, John Murray, London.

Davies, N.B. and Halliday, T.R. (1978) Deep croaks and fighting assessment in toads, *Bufo bufo*. *Nature*, **391**, 56–8.

Dawkins, R. and Krebs, J.R. (1978) Animals signals: Information or manipulation? in *Behavioural Ecology* (eds J.R. Krebs and N.B. Davies), pp. 282–309, Blackwell Scientific Publications, Oxford.

Diem, K.L. (1954) Use of a deer call as a means of locating deer fawns. *J. Wildl. Manage.*, **18**, 537–8.

Dominey, W.J. (1981) Maintenance of female mimicry as a reproductive strategy in bluegill sunfish (*Lepornis macrochinus*). *Environ. Biol. Fish.*, **6**, 59–64.

Drent, R. and Swierstra, P. (1977) Goose flocks and food finding: field experiments with barnacle geese in winter. *Wildfowl*, **28**, 15–20.

Dunham, D.W. (1966) Agonistic behaviour in captive Rose-breasted grosbeaks, *Pheuticus ludovicianus. Behaviour,* **27-28**, 160–73.

Dunn, G., Price, E.O. and Katz, L.S. (1987) Fostering calves by odor transfer. *Appl. Anim. Behav. Sci.,* **17**, 33–9.

Egan, J.P. (1978) *Signal Detection Theory and ROC Analysis,* Academic Press, New York.

Eibl-Eiblesfeldt, I. (1972) Similarities and differences in expressive movements, in *Non-verbal Communication* (ed. R.A. Hinde), pp. 297–314, Cambridge University Press, Cambridge.

Eisenbraun, E.J., Browne, C.E., Irvin-Willis, R.L. *et al.* (1980) Structure and stereochemistry of 4a,7,7a-nepetalactone from *Nepeta mussini* and its relationship to the 4a,7,7a- and 4a,7,7-nepetalactones from *N. cataria. J. Org. Chem.,* **45**, 3811–14.

Eisner, T. (1985) Still more on bird attacks. *New England. J. Med.,* **7**, 1232–3.

Ekman, P. and Friesen, W.V. (1969) The repertoire of non-verbal behaviour: categories, origins, usage and coding. *Semiotica,* **1**, 49–98.

Ekman, P. and Friesen, W.V. (1975) *Unmasking the Face,* Prentice-Hall, New Jersey.

Ellins, S.R., Catalano, S.M. and Schechinger, S.A. (1977) Conditioned taste aversion: a field application to coyote predation on sheep. *Behav. Biol.,* **20**, 91–5.

Enquist, M. (1985) Communication during aggressive interactions with particular reference to variation in choice of behaviour. *Anim. Behav.,* **33**, 1152–61.

Falls, J.B. (1982) Individual recognition by sounds in birds, in *Acoustic Communication in Birds, Vol. 12.* (eds D.E. Kroodsma and E.H. Miller), pp. 237–78, Academic Press, New York.

Falls, J.B. and Brooks, R.J. (1975) Individual recognition by song in white-throated sparrows. II. Effects of location. *Canad. J. Zool.,* **53**, 1412–20.

Feare, C.J., Greig-Smith, P.W. and Inglis, I.R. (1990) Current status and potential of non-lethal control for reducing bird damage in agriculture. *Proc. XIX International Ornithological Congress,* 493–506.

Fogel, B. (ed.) (1981) *Interrelations between People and Pets,* pp. 352, Charles C. Thomas, Springfield, Illinois, USA.

Fox, M.W. (1975) Pet-owner relationships, in *Pet Animals and Society* (ed. R.S. Anderson), pp. 37–53, Ballière Tindall, London.

Fox, M.W. and Cohen, J.A. (1977) Canid communication, in *How Animals Communicate* (ed. T.R.A. Sebeok), pp. 728–48, Indiana Univ. Press, Bloomington.

Fullard, W. and Reiling, A.M. (1976) An investigation of Lorenz's 'babyishness'. *Child Develop.,* **50**, 1191–3.

Galef, B.G. and Stein, M. (1985) Demonstrator influence on observer diet preference: Analysis of critical social interactions and olfactory signals. *Anim. Learn. Behav.*, **13**, 31–8.

Galef, B.G. and Wigmore, S.W. (1983) Transfer of information concerning distant foods: A laboratory investigation of the 'information centre' hypothesis. *Anim. Behav.*, **31**, 748–58.

Galef, B.G., Kennett, D.J. and Stein, M. (1985) Demonstrator influence on observer diet preference: Effects of simple exposure and the presence of a demonstrator. *Anim. Learn. Behav.*, **13**, 25–30.

Galef, B.G., Kennett, D.J. and Wigmore, S.W. (1984) Transfer of information concerning distant foods in rats: A robust phenomenon. *Anim. Learn. Behav.*, **12**, 292–6.

Galef, B.G., Mason, J.R., Preti, G. and Bean, N.J. (1988) Carbon disulfide: a semiochemical mediating socially-induced diet choice in rats. *Physiol. Behav.*, **42**, 119–24.

Garcia, J., Kovner, R. and Green, K.F. (1970) Cue properties vs palatability of flavors in avoidance learning. *Psychonom. Sci.*, **20**, 313–14.

Gardner, B.T. and Wallach, L. (1965) Shapes of figures identified as a baby's head. *Percept. Motor Skills*, **20**, 135–42.

Gawienowski, A.M., Orsulak, P.J., Stacewicz-Sapuntzakis, M. and Joseph, B.M. (1975) Presence of sex pheromone in preputial glands of male rats. *J. Endocr.*, **67**, 283–8.

Gawienowski, A.M., Orsulak, P.J., Stacewicz-Sapuntzakis, M. and Pratt, J.J. (1976) Attractant effect of female preputial gland extracts on the male rat. *Psychoneuroendocrinology.*, **1**, 411–18.

Gawienowski, A.M. and Stacewicz-Sapuntzakis, M. (1978) Attraction of rats to sulphur compounds. *Behav. Biol.*, **23**, 267–70.

Geist, V. (1971) *Mountain Sheep*, Univ. Chicago Press, Chicago.

Gier, H.T. (1975) Ecology and social behaviour of the coyote, in *The Wild Canids*, (ed. M.W. Fox), pp. 247–62, Van Nostrand Reinhold, Princetown, N.J.

Goffman, E. (1969) *Strategic Interaction*, University of Pennsylvania Press, Philadelphia.

Gorman, M.L. (1976) A mechanism for individual recognition by odour in *Herpestes auropunctatus. Anim. Behav.*, **24**, 141–5.

Gorman, M.L. (1986) The secretion of the temporal gland of the African elephant as an elephant repellent. *J. Trop. Ecol.*, **2**, 187–90.

Green, S. and Marler, P. (1979) The analysis of animal communication, in *Handbook of Behavioural Neurobiology*, Vol. 3 (eds P. Marler and J.G. Vendenbergh), pp. 75–158, Plenum Press, New York.

Greenwalt, C.H. (1968) *Bird Song, Acoustics and Physiology*, Smithsonian Inst. Pres., Washington DC.

Gustavson, C.R., Garcia, J., Hankins, W.G. and Rusiniak, K.W. (1974)

Coyote predation control by aversive conditioning. *Science*, **184**, 581–3.

Halliday, T.R. and Slater, P.J.B. (eds.) (1983) *Animal Behaviour, Vol. 2. Communication*, Blackwell Scientific Publications, Oxford.

Hamerstrom, F. (1957) The influence of a hawk's appetite on mobbing. *Condor*, **99**, 192–4.

Hamilton, W.J. (1965) Sun-orientated display of the Anna's humming-bird. *Wilson Bull.*, **77**, 38–44.

Hankins, W.G., Garcia, J. and Rusiniak, K.W. (1973) Dissociation of odor and taste in bait shyness. *Behav. Biol.*, **8**, 407–19.

Harvey, P.H. and Greenwood, P.J. (1978) Anti-predator defence strategies: some evolutionary problems, in *Behavioural Ecology* (eds J.R. Krebs and N.B. Davies), pp. 129–51, Blackwell Scientific Publications, Oxford.

Hepper, P.G. (1986) Kin recognition: functions and mechanisms, a review. *Biol. Rev.*, **61**, 63–93.

Hilgard, E.R. and Marquis, D.G. (1940) *Conditioning and Learning*, Appleton-Century-Crofts, New York.

Hillyer, D. (1976) An investigation using a synthetic porcine phero-mone and the effect on days from weaning to conception. *Vet. Rec.*, **98**, 93–4.

Hinde, R.A. (1954) Factors governing the changes in strength of a partially inborne response as shown by the mobbing behaviour of the chaffinch (*Fringilla coelebs*): II. The waning of the response. *Proc. Roy. Soc. B.*, **142**, 331–58.

Hinde, R.A. (1959) Behaviour and speciation in birds and lower verte-brates. *Biol. Rev.*, **34**, 85–128.

Hinde, R.A. (1961) Factors governing the changes in strength of a partially inborn response, as shown by the mobbing behaviour of the chaffinch (*Fringilla coelebs*): III. The interaction of short-term and long-term incremental and decremental effects. *Proc. Roy. Soc. B.*, **153**, 398–420.

Hinde, R.A. (1970) *Animal Behaviour*, McGraw-Hill, New York.

Hinde, R.A. (1981) Animal signals: ethological and game theory approaches are not incompatible. *Anim. Behav.*, **29**, 535–42.

Hold, B. and Schleidt, M. (1977) The importance of human odour in non-verbal communication. *Z. Tierpsychol.*, **43**, 225–48.

Hodge, W.H. (1946) Camels of the clouds. *Natn. Geographic. Mag.*, **89**, 641–56.

Horn, S.W. (1983) An evaluation of predatory suppression in coyotes using lithium chloride-induced illness. *J. Wildl. Manage.*, **47**, 999–1009.

Hulet, C.V. (1966) Behavioural, social and psychological factors affecting mating time and breeding efficiency in sheep. *J. Anim. Sci. Supple.*, **25**, 5–20.

Inglis, I.R. (1980) Visual bird scarers: an ethological approach, in *Bird Problems in Agriculture* (eds E.N. Wright, I.R. Inglis and C.J. Feare), pp. 121–43, British Crop Protection Council, Croydon, London.

Inglis, I.R. (1987) *Bird Scaring*, ADAS Leaflet P903, MAFF (Publications), Alnwick, Northumberland.

Inglis, I.R. and Isaacson, A.J. (1978) The responses of dark-bellied brent geese to models of geese in various postures. *Anim. Behav.*, **26**, 953–8.

Inglis, I.R. and Isaacson, A.J. (1984) The responses of woodpigeons (*Columba palumbus*) to pigeon decoys in various postures: a quest for super-normal alarm stimulus. *Behaviour*, **90**, 224–40.

Inglis, I.R. and Isaacson, A.J. (1987) Development of a simple scaring device for woodpigeons (*Columba palumbus*). *Crop Protect.*, **6**, 104–8.

Inglis, I.R., Huson, L.W., Marshall, M.B. and Neville, P.A. (1983) The feeding behaviour of starlings (*Sturnus vulgaris*) in the presence of 'eyes'. *Z. Tierpsychol*, **62**, 181–208.

Inglis, I.R., Fletcher, M.R., Feare, C.J. *et al.* (1982) The incidence of distress calling among British birds. *Ibis*, **124**, 351–5.

Jones, R.B. (1980) Reactions of male domestic chicks to two-dimensional eye-like shapes. *Anim. Behav.*, **28**, 212–18.

Kalat, J.W. (1977) Biological significance of food aversion learning, in *Food Aversion Learning* (eds N.W. Milgram, L. Kranes and T.M. Alloway), pp. 73–103, Plenum Press, New York.

Katcher, A.H. and Beck, A.M. (eds) (1983) *New Perspectives on Our Lives with Companion Animals*, University of Pennsylvania Press, Philadelphia.

Kerlinger, P. and Lehrer, P.H. (1982) Owl recognition and anti-predator behaviour of sharp-shinned hawks. *Z. Tierpsychol.*, **58**. 163–73.

Kling, J.W. and Stevenson, J.G. (1970) Habituation and extinction, in *Short-term Changes in Neural Activity and Behaviour* (eds G. Horn and R.A. Hinde), pp. 41–61, Cambridge Univ. Press, Cambridge.

Klopfer, P.H. and Gamble, J. (1966) Maternal 'imprinting' in goats: The role of chemical senses. *Z. Tierpsychol.*, **23**, 588–92.

Konig, C. (1968) Lautausserungen von Rauhfusskauz (*Aegolius funereus*) und Sperlingskauz (*Glaucidium passerium*). *Beihafte Vogelwelt*, **1**, 115–38.

Konishi, M. (1970) Evolution of design features in the coding of species-specificity. *Am. Zool.*, **10**, 67–72.

Krebs, J.R. (1974) Colonial nesting and social feeding as strategies for exploiting food reserves in the great blue heron (*Ardea herodias*).

Behaviour, **LI**, 99–134.

Kroodsma, D.E. (1977) Correlates of song organization among North American wrens. *Am. Nat.*, **111**, 995–1008.

Kruuk, H. (1976) The biological function of gulls' attraction towards predators. *Anim. Behav.*, **24**, 146–53.

Kruuk, H. and Sands, W.A. (1972) The aardwolf (*Proteles cristatus*) as a predator of termites. *E. Afr. Wildl. J.*, **10**, 221–7.

Lazarus, J. (1972) Natural selection and the functions of flocking in birds: a reply to Murton. *Ibis*, **114**, 556–8.

Lederer, E. (1950) Odeurs et parfums des animaux. *Fortschr. Chem. Org. Naturstaffe*, **6**, 87–153.

Lee, S. van der and Boot, L.M. (1955) Spontaneous pseudopregnancy in mice. *Acta. Physiol. Pharmacol. Neerl.*, **4**, 442–3.

Lee, S. van der and Boot, L.M. (1956) Spontaneous pseudopregnancy in mice II. *Acta Physiol. Pharmacol. Neerl.*, **5**, 213–14.

Lehrman, D. and Friedman, M. (1969) Auditory stimulation of ovarian activity in the Ring dove (*Streptopelia risoria*). *Anim. Behav.*, **17**, 494–7.

Levy, S.H., Levy, J.J. and Bishop, R.A. (1966) Use of tape recorded female quail calls during the breeding season. *J. Wildl. Manage.*, **30**, 426–8.

Lofting, H. (1922) *The Story of Dr Doolittle*, Jonathan Cape, London.

Lorenz, K. (1961) Phylogenetische Anpassung und adaptive Modifikation des Verhaltens. *Z. Tierpsychol.*, **2**, 1–29.

McConnell, P.B. and Baylis, J.R. (1985) Interspecific communication in cooperative herding: acoustic and visual signals from herding shepherds herding dogs. *Z. Tierpsychol.*, **67**, 302–28.

MacKay, D.M. (1972) Formal analysis of communicative processes, in *Non-verbal Communication* (ed. R.A. Hinde), pp. 3–25, Cambridge University Press, Cambridge.

Mackintosh, N.J. (1974) *The Psychology of Animal Learning*, Academic Press, London.

MacLeod, A.J. (1980) Chemistry of odours, in *Olfaction in Mammals* (ed. D.M. Stoddart), pp. 15–34, *Symp. Zool. Soc. Lond.*, No. 45, Academic Press, London.

MacNally, R.C. (1981) On the reproductive energetics of chorusing males: energy depletion profiles, restoration and growth in two sympatric species of Ranidella (Anura). *Oecologia*, **51**, 181–8.

MacNally, R.C. and Young, D. (1981) Song energetics of the bladder cicada, *Cystosoma saundersii*. *J. Exp. Biol.*, **90**, 185–96.

Magnus, D. (1958) Experimental analysis of some 'overoptimal' sign-stimuli in the mating behaviour of the fritillary butterfly *Argynnis paphia*. *Proc. 10th Int. Congr. Entomol.*, **2**, 405–18.

Mainardi, D., Marsan, M. and Pasquali, A. (1965) Causation of sexual preferences in the house mouse: The behaviour of mice reared by parents whose odour was artificially altered. *Soc. Ital. Sci. Nat. Museo Civico Storia Mat. Milano*, **104**, 825–38.

Marchington, J. (1977) *The Practical Wildfowler*, A & C Black, London.

Marler, P. (1959) Developments in the study of animal communication, in *Darwin's Biological Work* (ed. P.R. Bell), pp. 150–206, Cambridge Univ. Press, Cambridge.

Marler, P. (1967) Animal communication signals. *Science*, **157**, 769–74.

Marr, J.N. and Gardner, L.E. (1965) Early olfactory experience and later social behaviour in the rat: Preference, sexual responsiveness, and care of young. *J. Genet. Psychol.*, **107**, 167–74.

Martin, I.G. (1980) 'Homeochemic' intraspecific chemical signal. *J. Chem. Ecol.*, **6**, 517–19.

Maynard Smith, J.M. (1972) *On Evolution*, Edinburgh University Press, Edinburgh.

Maynard Smith, J.M. (1974) The theory of games and the evolution of animal conflicts. *J. Theor. Biol.*, **47**, 209–21.

Mech, L.D. (1977) Wolf-pack buffer zones as prey reservoirs. *Science*, **198**, 320–1.

Meese, G.B., Conner, D.J. and Baldwin, B.A. (1975) Ability of the pig to distinguish between conspecific urine samples using olfaction. *Physiol. Behav.*, **15**, 121–5.

Melchiors, A.M. and Leslie, C.A. (1985) Effectiveness of predator fecal odors as black-tailed deer repellents. *J. Wildl. Manage.*, **49**, 358–62.

Melrose, D.R., Reed, H.G.B. and Patterson, R.C.S. (1971) Androgen steroids associated with boar odour as an aid to the detection of oestrus in pig artificial insemination. *Br. Vet. J.*, **127**, 497–502.

Mertl, A.S. (1977) Habituation to territorial scent marks in the field by *Lemur catta. Behav. Biol.*, **21**, 500–7.

Messent, P.R. and Serpell, J.A. (1981) A historical and biological view of the pet-owner bond, in *Interrelations Between People and Pets* (ed. B. Fogle), pp. 5–22, Charles C. Thomas, Springfield, Illinois.

Meylan, A. and Murbach, R. (1966) Versuch zum Schutz einer Sonnenblumerpflanzung gegen Grunfinken *Carduelis chloris* mit Raubvogelattrappen. *Ornithologischer Beo Bachter*, **63**, 74–6.

Møller, A.P. (1988) False alarm calls as a means of resource usurpation in the Great Tit *Parus major. Ethology*, **79**, 25–30.

Monaghan, P. (1984) Applied ethology. *Anim. Behav.*, **32**, 908–15.

Moody, M.L. and Menzel, E.W. (1976) Vocalizations and their behavioural contexts in the tamarin *Saguinus fuscicollis. Folia Primatol.*, **25**, 73–94.

Morgan, P.A. and Howse, P.E. (1973) Avoidance conditioning of jackdaws (*Corvus monedula*) to distress calls. *Anim. Behav.*, **21**, 481–91.

Morgan, P.A. and Howse, P.E. (1974) Conditioning of jackdaws (*Corvus monedula*) to normal and modified distress calls. *Anim. Behav.*, **22**, 688–94.

Morris, D. (1956) The feather postures of birds and the problem of the origin of social signals. *Behaviour*, **9**, 75–113.

Morris, D. (1957) 'Typical intensity' and its relation to the problem of ritualisation. *Behaviour*, **11**, 1–12.

Morris, D. (1977) *Manwatching: a field guide to human behaviour*, Jonathan Cape, London.

Morse, M.A. and Balser, D.S. (1961) Fox calling as a hunting technique. *J. Wildl. Manage.*, **25**, 148–54.

Morton, E.S. (1977) On the occurrence and significance of motivation-structural rules in some bird and mammal sounds. *Am. Nat.*, **111**, 855–69.

Morton, E.S. (1986) Predictions from the ranging hypothesis for the evolution of long distance signals in birds. *Behaviour*, **99**, 65–86.

Moynihan, M. (1955) Some aspects of reproductive behaviour in the Black-headed Gull and related species. *Behaviour, Suppl., No. 4.*, 1–201.

Muller-Schwarze, D. (1974) Application of pheromones in mammals, in *Pheromones* (ed. M.C. Birch), pp. 452–4. North Holland, Amsterdam.

Muller-Schwarze, D., Altieri, R. and Porter, N. (1984) Alert odor from skin gland in deer. *J. Chem. Ecol.*, **10**, 1707–29.

Murray, B.G. and Gill, F.B. (1976) Behavioural interactions of blue-winged and golden-winged warblers. *Wilson Bull.*, **88**, 231–54.

Murton, R.K. (1974) The use of biological methods in the control of vertebrate pests, in *Biology in Pest and Disease Control* (eds D. Price-Jones and M.E. Solomon), Blackwell, Oxford.

Murton, R.K., Westwood, N.J. and Isaacson, A.J. (1974) A study of woodpigeon shooting: the exploitation of a natural animal population. *J. Appl. Ecol.*, **11**, 61–81.

Mykytowycz, R. (1965) Further observations on the territorial function and histology of the subcutaneous (chin) glands in the rabbit. *Anim. Behav.*, **13**, 400–12.

Nelson, J.B. (1969) The breeding behaviour of the Red-footed Booby *Sula sula. Ibis*, **111**, 357–85.

Parkes, A.S. and Bruce, H.M. (1962) Pregnancy-block in female mice placed in boxes soiled by males. *J. Reprod. Fertil.*, **4**, 303–8.

Perrone, M. and Paulson, D.R. (1979) Incidence of distress calls in mist-netting birds. *Condor*, **81**, 423–4.

Piersma, T. and Veen, J. (1988) An analysis of the communication function of attack calls in little gulls. *Anim. Behav.*, **36**, 773–9.

Ploog, W.D., Blitz, J. and Ploog, F. (1963) Studies on social and sexual behavior of the squirrel monkey (*Saimiri sciureus*). *Folia primat.*, **1**, 29–66.

Porter, R.H. and Moore, D. (1981) Human kin recognition by olfactory cues. *Physiol. behav.*, **27**, 493–5.

Power, D.M. (1966) Agonistic behaviour and vocalisations of orange-chinned parakeets in captivity. *Condor*, **68**, 562–81.

Prelog, V., Ruzicka, L., Meister, P. and Wieland, P. (1945) Steroide und Sexual-hormone. Untersuchungen uber den Zusammenhang zwischen Konstitution und Geruch bei Steroiden. *Helv. Chim. Acta.*, **28**, 618–28.

Raisbeck, G. (1963) *Information Theory*, MIT Press, Cambridge, Massachusetts.

Rand, W.R. and Rand, A.S. (1976) Agonistic behaviour in nesting iguanas: a stochastic analysis of dispute settlement dominated by the minimization of energy costs. *Z. Tierpsychol.*, **40**, 279–99.

Rhijn, J.G. van and Vodegel, R. (1980) Being honest about one's intentions: an evolutionary stable strategy for animal conflicts. *J. Theor. Biol.*, **85**, 623–41.

Richard, D.B. and Thompson, N.S. (1978) Critical properties of the assembly call of the common American crow. *Behaviour*, **64**, 184–203.

Richards, D.G. (1981) Alerting and message components in songs of Rufous-sided Towbees. *Behaviour*, **76**, 223–49.

Ritter, F.J., Bruggemann, I.E.M., Gut, J. and Persoons, C.J. (1982) Recent pheromone research in the Netherlands on muskrats and some insects, *A.C.S. Symposium Series No. 190*, Insect Pheromone Technology: Chemistry and Application, pp. 107–30.

Rochelle, J.A., Gaudity, I., Oita, K. and Oh, J. (1974) New developments in big game repellents, in *Proceedings of a Symposium on Wildlife and Forest Management in the Pacific North West*, pp. 103–112, Oregon State University, Corvallis.

Rohwer, S. (1977) Status signalling in Harris sparrows; some experiments in deception. *Behaviour*, **61**, 107–29.

Russell, E.S. (1943) Perceptual and sensory signs in instinctive behaviour. *Proc. Linn. Soc. Lond.*, **154**, 195–216.

Rydén, O. (1978) Differential responsiveness of great tit nestlings, *Parus major*, to natural auditory stimuli. *Z. Tierpsychol.*, **47**, 236–53.

Sakai, T., Nakajima, J.K., Yoshihara, K. and Sakan, T. (1980) Revisions of the absolute configurations of C-8 methyl groups in dehydroirid, neonepetalactone and matatabiether from *Actinidia polygama. Tetrahedron*, **36**, 3115–119.

Scaife, M. (1976) The response to eye-like shapes by birds. II. The importance of staring, pairedness and shape. *Anim. Behav.*, **24**, 200–6.

Schultz, E.F. and Tapp, J.T. (1973) Olfactory control of behavior in rodents. *Psychol. Bull.*, **79**, 21–44.

Sebeok, T.A. (ed.) (1968) *Animal Communication: Techniques of Study and Results of Research*, Indiana University Press, Bloomington.

Sebeok, T.A. (ed.) (1977) *How Animals Communicate*, Indiana University Press, Bloomington.

Seton, E.T. (1927) *Lives of Game Animals*, Vol. III, Doubleday, New York.

Schemnitz, S.D. (1980) *Wildlife Management Technique Manual*, The Wildlife Society, Washington, DC.

Shalter, M.D. (1975) Lack of spatial generalization in habituation tests of fowl. *J. Comp. Physiol. Psychol.*, **89**, 258–62.

Shalter, M.D. (1978) Effect of spatial context on the mobbing reaction of pied flycatchers to a predator model. *Anim. Behav.*, **26**, 1219–22.

Shiovitz, K.A. (1975) The process of species-specific song recognition by the Indigo Bunting and its relationship to the organization of avian acoustical behaviour. *Behaviour*, **55**, 128–79.

Shorey, H.H. (1976) Pheromones, in *How Animals Communicate* (ed. T.A. Sebeok), pp. 137–63, Indiana University Press, Bloomington.

Shumake, S.A. (1977) The search for applications of chemical signals in wildlife management, in *Chemical Signals in the Vertebrates* (eds D. Muller-Schwartze and M.M. Mozell), pp. 359–76, Plenum Press, New York.

Signoret, J.P. (1970) Reproductive behaviour in pigs. *J. Reprod. Fert.* (Suppl.), **11**, 105–17.

Singh, P.B., Brown, R.E. and Roser, B. (1987) MHC antigens in urine as olfactory recognition cues. *Nature*, **327**, 161–4.

Slater, P.J.B. (1973) Describing sequences of behaviour, in *Perspectives in Ethology*, (eds P.P.G. Bateson and P.H. Klopter), pp. 131–53, Plenum Press, New York.

Slater, P.J.B. (1983) The study of communication, in *Animal Behaviour. Vol. 2. Communication* (eds T.R. Halliday and P.J.B. Slater), pp. 9–42, Blackwell Scientific Publications, Oxford.

Smith, W.J. (1968) Message meaning analysis, in *Animal Communication* (ed. T.A. Sebeok), pp. 44–60, Indiana University Press, Bloomington.

Smith, W.J. (1977) *The Behaviour of Communicating*, Harvard

University Press, Cambridge, Mass.

Smuts, G.L., Whyte, I.J. and Dearlove, T.W. (1977) Advances on the mass capture of lions (*Panthera leo*). *Int. Congr. Game Biol.*, **13**, 420–31.

Stacewicz-Sapuntzakis, M. and Gawienowski, A.M. (1977) Rat olfactory response to aliphatic acetates. *J. Chem. Ecol.*, **3**, 411–17.

Stager, K.E. (1964) The role of olfaction in food location by the turkey vulture (*Cathartes aura*). *LA County Museum Contributions in Science*, No. 81.

Stevens, D.A. and Gerzoy-Thomas, D.A. (1977) Fright reactions in rats to conspecific tissue. *Physiol. Behav.*, **18**, 47–51.

Stirling, I. and Bendrell, J.F. (1966) Census of blue grouse with recorded calls of a female. *J. Wildl. Manage.*, **30**, 184–7.

Stoddart, D.M. (1976) Effect of the odour of weasels (*Mustela nivalis* L.) on trapped samples of their prey. *Oecologia*, **22**, 439–41.

Stoddart, D.M. (1980) *The Ecology of Vertebrate Olfaction*, Chapman and Hall, London.

Stokes, A.W. (1962) Agonistic behaviour among blue tits at a winter feeding station. *Behaviour*, **19**, 118–38.

Sullivan, T.P. and Crump, D.R. (1984) Influence of mustelid scent-gland compounds on suppression of feeding by snowshoe hares (*Lepus americanus*). *J. Chem. Ecol.*, **10**, 1809–21.

Sullivan, T.P., Nordstrom, L.O. and Sullivan, D.S. (1985a) Use of predator odors as repellents to reduce feeding damage by herbivores. I. Snowshoe Hares (*Lepus americanus*). *J. Chem. Ecol.*, **11**, 903–19.

Sullivan, T.P., Nordstrom, L.O. and Sullivan, D.S. (1985b) Use of predator odors as repellents to reduce feeding damage by herbivores. II. Black-tailed deer (*Odocoileus hemionus columbianus*). *J. Chem. Ecol.*, **11**, 921–35.

Summers, R.W. (1985) The effect of scarers on the presence of starlings (*Sturnus vulgaris*) in cherry orchards. *Crop Protect.*, **4**, 520–8.

Tansley, K. (1965) *Vision in Vertebrates*, Chapman and Hall, London.

Tester, A.L. (1963) Olfaction, gustation and the chemical sense in sharks, in *Sharks and Survival* (ed. P.W. Gilbert), pp. 255–81, D.C. Heath and Co., Boston, Mass.

Theimer, E.T. and Davies, J.T. (1967) Olfaction, musk odour and molecular properties. *J. Agric. Food Chem.*, **15**, 6–14.

Thompson, R.F. and Spencer, W.A. (1966) Habituation: A model phenomenon for the study of the neural substrates of behaviour. *Psychol. Rev.*, **73**, 16–43.

Thompson, T. (1969) Conditioned avoidance of the mobbing call by chaffinches. *Anim. Behav.*, **17**, 517–22.

Timms, A.M. and Kleerekoper, H. (1972) The locomotor response of male *Ictalurus punctatus*, the channel catfish, to a pheromone released by the ripe female of the species. *Trans. Am. Fish. Soc.*, **101**, 302–10.

Tinbergen, N. (1942) An objective study of the innate behaviour of animals. *Biblioth Biother*, **1**, 39–98.

Tinbergen, N. (1952) Derived activities: their causation, biological significance, origin and emancipation during evolution. *Quart. Rev. Biol.*, **27**, 1–32.

Tinbergen, N. (1953) *The Herring Gull's World*, Clarendon Press, Oxford.

Tinbergen, N. (1959) Comparative studies on the behaviour of gulls (*Laridae*): a progress report. *Behaviour*, **15**, 1–70.

Tinbergen, N. and Perdeck, A.C. (1950) On the stimulus situation releasing the begging response in the newly hatched herring gull chick (*Larus argentatus argentatus*). *Behaviour*, **3**, 1–39.

Todd, N.B. (1962) Inheritance of the catnip response in domestic cats. *J. Hered.*, **53**, 54–6.

Todd, N.B. (1963) The Catnip Response, PhD Thesis, Harvard University.

Vines, G. (1981) Wolves in dog's clothing. *New Scientist*, **10**, 648–52.

Voith, V.L. (1981) Attachment between people and their pets: behaviour problems of pets that arise from the relationships between pets and people, in *Interrelations between People and Pets* (ed. B. Fogle), pp. 271–94, Charles C. Thomas, Springfield, Illinois.

Walther, F.F. (1984) *Communication and Expression in Hoofed Mammals*, Indiana University Press, Bloomington.

Watson, A. and Miller, G.R. (1971) Territory size and aggression in a fluctuating red grouse population. *J. Anim. Ecol.*, **40**, 367–83.

Whitten, W.K. (1959) Occurrence of anoestrus in mice caged in groups. *J. Endocrinol.*, **18**, 102–7.

Wiley, R.H. and Richards, D.G. (1978) Physical constraints on acoustic communication in the atmosphere: Implications for the evolution of animal vocalisations. *Behav. Ecol. Sociobiol.*, **3**, 69–94.

Wiley, R.H. and Richards, D.G. (1982) Adaptations for acoustic communication in birds: Sound transmission and signal detection, in *Acoustic Communication in Birds Vol. 1* (eds D.E. Kroodsma and E.H. Miller), pp. 131–81, Academic Press, New York.

Wood, T. (1978) Cycloalkylation of aromatics with isoprene. Musks to herbicides. *Perfum and Flavor*, **3**, 33.

Part Two
Improving Productivity

5

Aquaculture

NEIL B. METCALFE

There has been a massive expansion in aquaculture in the past twenty years, partly as a consequence of decline in many traditional fisheries and partly due to improvements in fish husbandry and disease control making fish farming a viable industry. Unlike conventional agriculture, there is usually no established tradition of rearing the animals in question, so that the husbandry practices taken for granted by land-based farmers have had to be learnt rapidly through trial and error by their fish-farming counterparts. This process is made more difficult by the nature of the environment: a dairyman can observe known individual cows to see whether (and on what) they are feeding, whereas salmon farmers catch only glimpses of fish (which will be disturbed by their presence). Far from recognizing individuals, the fish farmer may have only a rough idea of how many fish are present. Moreover, it is rarely possible to see how much of the food is eaten, since it either disintegrates or is washed away. Knowledge of social interactions is similarly limited. In these circumstances ethologists and behavioural ecologists have an especially important role to play, since the information they specialize in providing is a unique record of what is really happening below the water surface.

This chapter concentrates on the ways in which studies of behaviour have (or could) contribute to aquaculture. Some of the approaches have yet to be taken up by the industry, partly because during the early stages of its expansion biological interest focused more on the nutritional and veterinary problems. Now that many of these have been solved, there is a slow realization that further improvements in husbandry may depend on knowing more about how the animals behave. I have concentrated on the farming of fish, and on salmonid fish in particular, since these are economically some of the most important species in the western world where the majority of the behavioural research has been carried out. However, many of the principles discussed apply equally well to other systems.

5.1 REARING FISH

5.1.1 What should they be fed on?

It is rarely feasible to feed fish farmed in an intensive system on a natural diet; the diet of commercial species ranges from phytoplankton through insects to other fish, and in most cases there are insurmountable problems in either obtaining enough of the food, storing it in a stable and compact form or devising an automatic method of dispensing it. Most of the examples of using natural foods are low-intensity (or 'extensive') systems where the farmer creates a 'mini-ecosystem', such as the ponds used to produce milkfish (*Chanos chanos*) in the tropics, where natural production (sometimes boosted by fertilizer) of phytoplankton and algae is sufficient to sustain the relatively low densities of fish (Pitcher and Hart, 1982).

In the majority of cases, therefore, the food must be manufactured. Extensive research into the nutritional requirements of fish has identified the optimal composition of feed for many species (Halver and Tiews, 1979; Cowey *et al.*, 1985). The problem for the behavioural scientist is in finding a method of packaging that produces the best feeding response from the fish — there is little point in making a food that contains the ideal components if the fish are physically unable to eat it or do not even recognize it as food. (Similar problems are encountered in land-based agriculture — see the discussion of food recognition by lambs in Chapter 6.)

Fish tend to be rather selective in what they feed on in the wild; when given a choice they (perhaps not surprisingly) prefer the more profitable prey items (Werner and Hall, 1974; Ringler, 1979). However, theories of optimal foraging have highlighted the fact that profitability is not just measured in terms of the energy or nutrient content of the food: the rate at which different prey types are encountered and the time taken to manipulate and swallow food items once discovered ('handling times') are also crucial in determining the optimal diet (Krebs *et al.*, 1983, and Chapter 2). By adequate provisioning the fish farmer can ensure that the encounter rates remain high, but must study feeding behaviour to get an idea of handling times.

Most species of fish are unable to bite off mouthfuls of food, so that handling times increase markedly as the size of food items approaches the mouth width, and larger items cannot be swallowed at all. For instance, Werner (1974) found that the time taken by bluegill sunfish (*Lepomis macrochirus*) to handle individual *Daphnia* increased from less than a second to over 80s as the size of the *Daphnia* approached the diameter of the mouth. An interesting consequence of this is that the largest food pellets that a fish can swallow are not the most profitable. Wankowski (1979, 1981) found that juvenile Atlantic salmon (*Salmo salar*) quite often made an attempt to eat these large pellets, but usually quickly rejected them, since they would gain more in terms of energy per unit time by concentrating on smaller ones. One problem with small prey items is that they are harder to detect. This is especially important for species (such as juvenile salmonids) that are sit-and-wait predators in fast flowing water, darting out to intercept items as they are carried past in the drift. The distance over which small items can be seen is short, so that many may be missed (Wankowski, 1981; Dunbrack and Dill, 1983). The optimum is therefore a compromise. Wankowski (1979, 1981) found that salmon showed the greatest behavioural response to pellets with a diameter approximately 25% of the mouth breadth, and these behavioural data were confirmed by subsequent growth trials. Mouth breadth is not the easiest thing to measure, but luckily it increases linearly with fish length, and so the farmer can judge the correct pellet size by the length of the fish. A similar approach has been used to determine optimal pellet sizes for farmed eels (*Anguilla anguilla*) (Knights, 1983).

The shape, texture and colour of pellets has often been determined by the economics of manufacture rather than by any preference of the fish. The easiest shape of small pellet to produce is a roughly spherical 'crumb'. However, salmon have been found to be far less likely to approach and attack crumb compared with rod-shaped pellets composed of the same material, possibly because most of their natural food is elongate and a round object is more likely to be something inedible, such as grit (Stradmeyer *et al.*, 1988). Fish also demonstrate clear preferences when given a choice between food types differing only in colour. Naive rainbow trout (*Oncorhynchus mykiss*) and Atlantic salmon have both been found to prefer red items (Ginetz and Larkin, 1973; Clarke and Sutterlin, 1985), although this innate preference can be altered by previous experience, the fish then preferring the colour they have been reared on (Figure 5.1). Jakobsen *et al.* (1987) found that farmed salmon grew faster when fed on two colours of pellet simultaneously than when fed on either colour alone. They suggest that normally the fish incur a 'confusion cost' when feeding (Milinski, 1984), being distracted by the presence of many alternative food items; the confusion is reduced when the diet is more varied. There was some indication that the smallest fish benefited most from this reduction in confusion, possibly because they must avoid the larger fish and so cannot devote as much attention to feeding. There is also evidence that the commercial salmon pellets commonly used

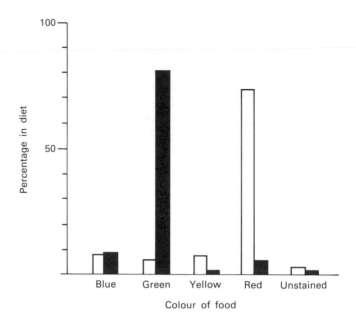

Figure 5.1 The colour of dyed food preferred by naive Atlantic salmon fry (white bars) and fry given a previous meal of green food (black bars). They have an innate preference for red, but this can quickly be overridden by experience. (After Clarke and Sutterlin, 1985).

by farmers are harder in texture than those that produce an optimal response from the fish (Stradmeyer *et al.*, 1988). However, while colour is relatively easy to manipulate, the increased costs of producing and storing pellets of the preferred shape and texture may prevent them being economically viable except perhaps for use at critical times of the life cycle when growth must be maximized.

5.1.2 How should the food be dispensed?

Stream-living fish (i.e. most juvenile salmonids) are adapted to select the most profitable position in which to wait for food items; this is usually in, or adjacent to, fast flowing water (Smith and Li, 1983; Fausch, 1984). They may orientate and respond only to food items that are moving (Rimmer and Power, 1978). The farmer must there-fore arrange the food supply within the tank or cage so that the pellets are kept in motion for a long as possible; the simplest solution is to arrange for food to be dispensed at regular intervals into the water currents entering the tank. For larger fish it may be possible to use feeders that the fish themselves operate: the fish trigger the release of food pellets whenever they hit or brush against a stiff wire suspended in the water, and so learn to feed themselves effectively by operant conditioning. The design of such feeding regimens draws heavily on animal learning theory. The potential advantage of 'demand feeders' is that the fish only activate the feeders when hungry, so reducing both wastage and pollution. This may be especially important in large ponds or cages where the farmer is unable to monitor mortality through disease or predation, and so has little idea of the number of fish present and hence amount of food required. The principle has been tried on a variety of systems, with mixed success. Landless (1976) found that, while isolated rainbow trout could learn to use the feeders, keeping them in groups of 25 resulted in only one member of the group learning to operate the mechanism, the rest feeding on the pellets it released. There is also a danger with this approach that the fish will feed at a maximum rate, which is not the rate that maximizes the efficiency of digestion (i.e. weight gained per gram eaten) (Elliott, 1982). This is an example of the animal's optimization of costs and benefits being potentially incompatible with that of the farmer, since the latter is more interested in maximizing efficiency of growth rather than growth rate *per se*. Nonetheless, Tipping *et al.* (1986) found that using demand feeders in place of conventional automatic dispensers on five acre ponds of steelhead trout (*Oncorhynchus mykiss*) led to much higher feeding efficiencies, with no reduction in survival rates or growth (Table 5.1).

An advantage of rearing fish in large open systems such as lakes is that they can gain much of their food from natural sources, and may only need to be given a supplement. The problem then arises as to how to dispense a relatively small amount of rapidly-degrading food efficiently into a large area. Training the fish to gather at a localized

Table 5.1 A comparison of survival rates and feed conversion efficiencies (weight of food used/weight gained by surviving fish) for pond-reared steelhead trout fed by demand feeder or automatic blower feeder (After Tipping *et al.*, 1986)

Feeder type	Survival (percent ± s.e.)	Feed conversion (± s.e.)
Demand feeders	69.6 ± 2.4	1.87 ± 0.03
Blower feeders	50.2 ± 5.3	3.20 ± 0.77

food source at feeding times increases the likelihood of the food being consumed, although it may create localized pollution due to the build-up of uneaten food (Gowan and Bradbury, 1987). The times of feeding can be signalled by broadcasting the equivalent of a dinner-gong: a pond population of 13,000 rainbow trout learnt within 10 days to associate a 150Hz pure tone with feeding, 50–90% of them gathering within a minute of hearing it in the vicinity of the feeder (Abbott, 1972). When Phillips (1982) extended the scale of this approach to an 8-acre lake, he found that it became harder to train the fish: not all of the rainbow trout would reach the food in time when the sound stimulus and food were presented. This results in only partial reinforcement of the fishes' response, and so may lead to response extinction. However, the mere sight of feeding fish attracted others towards the feeder, so that fish exploited the food source even in the absence of a sound signal.

5.1.3 What determines appetite and growth?

The poikilothermic nature of fish results in their metabolic processes being highly constrained by ambient temperature; this therefore tends to determine maximal food intake rates, feeding efficiency and growth rates (Elliot, 1982). The fish farmer can therefore determine feeding regimens purely on the basis of the biomass of fish present and the water temperature. However, there is evidence of seasonal endogenous rhythms in metabolism, related to particular life-history strategies, that may act independently of temperature. One well-known example concerns the loss of appetite in adult salmon as they become sexually mature; the salmon stop feeding when they return to freshwater, despite the presence of suitable food items. A second, more complex, phenomenon is found in juvenile Atlantic salmon before they have left freshwater. These fish undertake the seaward migration (as smolts) in the spring, but at a variable age. Hatcheries generally produce both 1- and 2-year old smolts. It is possible to recognize those fish that will become 1-year old smolts in their first autumn, since the size frequency distribution of fish within a single tank suddenly changes from a normal distribution to a bimodal one. The larger individuals, which form the Upper Modal Group (UMG), will become 1-year old smolts, while the smaller salmon in the Lower Modal Group (LMG) will not reach the migratory stage for a further year (Thorpe, 1977). The LMG fish are clearly less profitable for the farmer, since they must be fed for an extra year during which they occupy valuable tank space on the farm.

The reason for the development of a bimodal distribution is that the

LMG fish start to lose appetite from mid-summer onwards, so that by late autumn they are only eating enough to stay alive. This natural anorexia is independent of both temperature and food abundance (Metcalfe *et al.*, 1986), and persists until the following spring. Meanwhile, their UMG siblings, which are preparing to smolt, maintain their appetite (and even show an appetite surge in the autumn, despite falling temperatures (Metcalfe *et al.*, 1988)), so that the two groups of fish soon become separable on the basis of size. It appears that these development patterns are more environmentally than genetically determined; all fish seem to have the capability to become 1-year old smolts, but in the middle of their first summer some fish adopt a strategy that leads to delayed smolting. The developmental pathway that is followed is influenced by early growth conditions, so that if conditions are favourable for growth a greater proportion of the population will adopt the UMG strategy (Thorpe, 1987).

5.1.4 How might social behaviour constrain feeding and growth?

Fish farmers are keen to rear fish at high densities, partly because the initial costs of buying the fish are small when compared with the costs of the tanks, ponds or cages in which they are to be kept. It would be surprising if this did not cause the same kinds of behavioural problems found in other high-intensity food production units such as farms with intensive husbandry systems. In trying to find ways around these, we must first understand the normal social behaviour and spacing patterns of the animals.

Many salmonids are territorial when young, aggressively defending stable feeding territories on the stream bed. This could potentially affect the aquaculture industry, since in the wild territoriality may set an upper limit on population density (Slaney and Northcote, 1974, and see Figure 3.2 this volume). The degree of aggression shown by juvenile salmonids is both genetically and environmentally mediated. In an elegant series of experiments relating behaviour to both genetics and life-histories, Ferguson and Noakes (1982) showed that juvenile brook charr (*Salvelinus fontinalis*) were inherently more aggressive than the closely-related juvenile lake charr (*S. namaycush*). As their names suggest, the two species are found in different habitats, the former normally holding territories in streams while the latter feeds pelagically in lakes. Under standardized conditions brook charr tend to be aggressive and rather stationary, whereas lake charr are the opposite. Moreover, their F1 hybrids and backcrosses are intermediate (Figure 5.2). An analogous intraspecific example is the chinook salmon

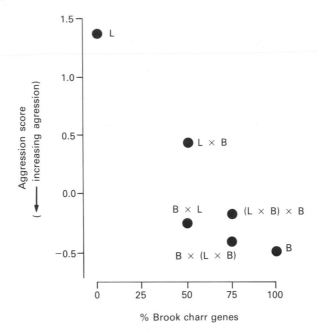

Figure 5.2 The mean aggressiveness scores of lake (L) and brook (B) charr and their hybrids is correlated with the genetic composition of the fish (represented as the percentage of genes derived from brook charr). The female parent is given first in the hybrids. (After Noakes and Grant, 1986).

(*Oncorhynchus tshawytscha*), which may live for several years in streams (stream-type populations) or migrate to sea within months of emerging (ocean-type). Laboratory tests on newly-emerged fry of the two forms showed that stream-type fish were inherently far more aggressive, and so better adapted to a territorial existence (Taylor and Larkin, 1986; Taylor, 1988).

On top of this genetic influence is also an environmental one, giving a degree of plasticity; the fish will (to some extent) adapt their degree of aggressiveness to environmental conditions. Territoriality can be viewed as a strategy that is only worth adopting when the benefits outweigh the costs (Davies and Houston, 1984; see also Chapter 3). It should therefore break down when fish densities or water currents increase above a certain limit (i.e. when the costs of defending it against intruders are too high), or when food levels are either very low or very high (i.e. when there is no real benefit to be gained in defending an area). The data on field and laboratory studies of dispersion in salmonids tend to support these predictions (Noakes and

Grant, 1986), the fish tending to be territorial (rather than schooling) only at intermediate food levels and lower population densities. Since there is this natural flexibility, fish reared at high densities will usually not attempt to be territorial, and will not necessarily suffer because of it (although Ferno and Holm (1986) noticed that a few individual salmon parr were able to hold territories and monopolize large areas in tanks containing several thousand fish).

While territoriality is a strategy that may be turned on or off (and so can be manipulated), dominance relationships are harder to eliminate. Fish form dominance hierarchies even when in schools, and the most dominant individuals obtain the best feeding positions (Fausch, 1984). (Similar results are obtained in farmed piglets — see Chapter 6.) This results in poorer growth by subordinates, either because they must remain in less profitable feeding sites (Fausch, 1984; Metcalfe, 1986) or because their appetite is suppressed. Koebele (1985) found that even the sight of a dominant was enough to reduce the food intake of subordinate cichlids (*Tilapia zillii*), despite their being given excess food. The variation in growth leads to a skewed size distribution, with relatively few large (dominant) fish and a high proportion of small subordinates; moreover the smaller fish tend to have a lower condition index (weight/length^3). This phenomenon of 'growth depensation' in subordinates has been found in a variety of commercially-reared species including Arctic charr (*Salvelinus alpinus*) and *Tilapia* (Jobling, 1985; Koebele, 1985; Saclauso, 1985), although it is not universal and clearly depends on the natural social structure of the species concerned. For example, small cod (*Gadus morhua*) do not experience suppressed growth when kept caged at high densities with larger individuals, possibly because cod are non-territorial and relatively unaggressive in the wild (Jobling, 1985).

As well as a reduced growth rate, subordinate fish may suffer from physical injuries. A high proportion of reared fish have damaged caudal and dorsal fins. While this is often thought to be due to abrasion against the tank walls or cage netting, Abbott and Dill (1985) concluded that it was due to aggression, since these were the areas of the body receiving most bites in steelhead trout. The experience of being subordinate can also lead to physiological stress, as found in eels (Peters *et al.*, 1980), although this may only be important when the animals are closely confined. A combination of greater risk of injuries, increased stress and poorer feeding rates may result in subordinates having greater mortality rates, as found by Yamagishi (1962) in rainbow trout.

In addition to these overt responses to dominance status, there may be more subtle effects. In juvenile Atlantic salmon, there is initially no correlation between status and size in hatchery populations: dominant fish are no bigger than subordinates. However, dominants are more likely to adopt the fast development strategy and join the UMG than are subordinates of the same size: the more dominant a fish is, the greater the probability of it becoming a 1- rather than 2-year old smolt (Figure 5.3). This appears to be related to differences in competitive ability. It has been shown experimentally that those salmon that have joined the LMG tend to respond to the presence of a feeding competitor (simulated by a mirror) by aborting attacks on food; they will approach food pellets but often pull away if they see a competing fish heading for the same item (Metcalfe, 1989). This timidity will result in lowered feeding rates when contesting for food with other fish, and so may be the cause of these fish subsequently following a life-history strategy of slower development and deferred smolting. In contrast, UMG fish are unaffected by the sight of a competitor, and so are more likely to maintain food intake regardless of the density of other fish (Metcalfe, 1989). The greater the competition for food or feeding stations, the greater the subordinate fish are disadvantaged and thus the more likely it will be that they delay smolting.

While these various effects of dominance may be found in natural wild populations, they may be exacerbated in an aquacultural setting, since subordinate fish do not have the option of evading dominants. Instead, we must aim to minimize the stress imposed on subordinates by the continual presence of high densities of more dominant fish. Paradoxically, one solution is to increase the numbers of fish per tank still further, since this will lead to a reduced polarization of the social structure. Thus Yamagishi (1962) found that increasing the size of a population of rainbow trout from 6 to 120 resulted in lower growth depensation and lower mortality, due to the collapse of the dominance hierarchy. In Yamagishi's experiment the density was kept constant, by moving the fish to larger and larger containers. Densities can be increased to surprisingly high levels without apparent effect, but there are critical limits, dependent on the species, beyond which growth declines and mortality increases regardless of feeding rate (Refstie and Kittelsen, 1976).

Fish farmers often attempt to overcome the effects of growth depensation by size-grading their fish. In effect, the fish are 'sieved' (often using semi-automated graders) into different size categories, which are then reared separately. The rationale is that removal of the larger fish releases the smaller from suppression, so leading to more even growth. However, this is based more on faith than fact, and experiments by Jobling and Reinsnes (1987) demonstrate that the

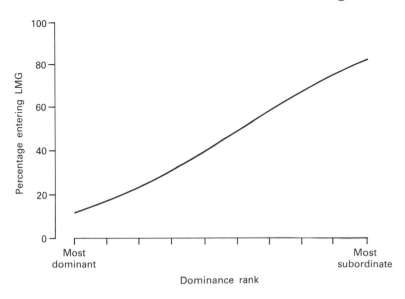

Figure 5.3 The percentage of Atlantic salmon entering the Lower Modal Group (and so delaying smolting for a year) in relation to dominance rank for fish in the middle of the size distribution two months after first feeding (After Metcalfe *et al.*, 1989.)

benefits of this time-consuming and stressful practice are questionable. They took two hatchery populations of young Arctic charr and graded one into two size categories, keeping the other as an ungraded control. The populations were then reared separately, keeping all tanks at the same density. Sorting did indeed result in some of the smaller fish growing faster, but this was offset by a slowed growth of some of the fish in the 'larger' group. As a result, five months after sorting, the total biomass of the combined sorted tanks was indistinguishable from that of the unsorted control. The probable explanation is that dominants and subordinates will arise out of any population, whether the fish are large or small, and so segregating them by size will have little long-term effect, especially if dominance is not correlated with size (Metcalfe *et al.*, 1989). Moreover, categorizing the fish according to size may actually increase competition, since all fish within a tank will then be competing for the same narrow size range of optimal food items (see earlier section concerning optimal diet choice).

So what can be done to minimize the effects of dominance? Since most of the aggression observed in farmed fish is in the form of contests over space or food, increasing the availability of these resources should reduce the competition. One of the simplest ways of making it harder for dominants to monopolize the food is to increase the feeding frequency. Jobling (1983) found that the largest Arctic charr in groups fed once every two days grew as fast as their counterparts in groups fed twice a day, but the smallest did much worse, leading to greater variation in growth (Figure 5.4). It is important to note that these effects occurred even though the fish were fed to apparent satiation each time (i.e. until no more food was taken). This suggests that the poorly-growing fish in the group given restricted access to food were intimidated from feeding by the presence of feeding competitors, and so did not attempt to feed when these had become satiated. The spatial and temporal availability of the food can also be increased by presenting it in the form of floating pellets. These have the advantage over sinking pellets in that they spread over the surface (and so are less localized), and do not disappear through the bottom of cages or into the sediments of ponds. Pond-reared channel catfish (*Ictalurus punctatus*) showed much less variation in size (and hence less growth depensation) when fed on floating rather than sinking pellets (Konikoff and Lewis, 1974).

Increasing the availability of space per fish need not necessarily mean changing the overall density of fish per tank. Many farmed species have specific micro-habitat preferences that result in their not using all the available space equally. For instance, Atlantic salmon parr make use of overhead cover when this is available, possibly as a means of reducing their conspicuousness to avian predators. In open tanks they tend to avoid the most exposed areas, clustering around the edge and near structural features such as drain holes. They can be made to spread out more evenly simply by providing a sheet of overhead cover, suspended low over the tank. Pickering *et al.* (1987), following up an earlier experiment by Thorpe and Wankowski (1979), found that provision of this cover led to an increase both in growth rate and in the percentage of salmon that smolted after only one year (from 22% to 38%); the fish with cover also showed fewer signs of haematological stress (i.e. the loss of thrombocytes and lymphocytes from the blood).

Another approach is to rear fish in mixed-species groups, making use of the fact that some salmonid species hold station in the water column while others do so on the bottom. The former species are less dependent on cover, so that their growth does not improve when cover is added; (Pickering *et al.*, 1987). Mork (1982) found that fish in mixed groups of Atlantic salmon and either rainbow or sea trout (*Salmo trutta*) grew better than when reared as monocultures of similar total density; the trout tended to swim in the upper half of the

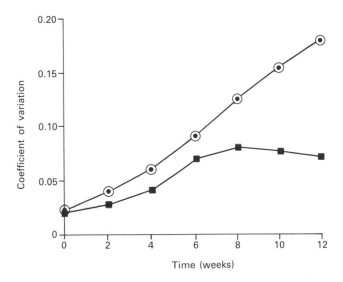

Figure 5.4 Changes in the coefficient of variation (variance/mean^2) of arctic charr weights with time since starting a feeding regimen of one meal every two days (upper line) or two meals a day (lower line). (After Jobling, 1983.)

water column while the salmon occupied the bottom, so that the competition for space was reduced. The second species may also act as a form of overhead cover — salmon kept with arctic charr (which are pelagic) tend to swim much higher in the water column than when reared alone (Jens Chr. Holm, personal communication). A drawback of duoculture is that the different species may not grow at the same rate, so that periodically the populations may have to be re-sorted in order to reduce the size variation (and hence required size range of food pellets) per tank.

A more equitable distribution of fish in rearing tanks can also be achieved by modifying the design of the tank. The standard design for rearing young salmonids is based more on the ease of manufacture and of day-to-day maintenance than on the behavioural requirements of the fish. The water supply usually enters tangentially to the side of the tank, creating a circular current that spirals in to the drain set in the centre. The tanks are often square with rounded corners, and produce local current eddies at the corners and around the drain. The food, carried on the current, is thus unevenly distributed in the tank, so that some fish are able to obtain much better feeding stations than others. Thorpe and Wankowski (1979) developed a different style of tank to try and overcome this variability. Water enters their circular tank at the centre, and flows outwards radially in all directions across the tank floor. A ring of inward-facing fish forms around the centre; their position can be manipulated by providing a ring of cover suspended just above the tank floor. This design has the advantage that all fish can maintain station under cover and all have much the same access to the food, which enters with the water. Growth rates of fish reared in these radial flow tanks tends to be less variable than in conventional tanks, but their drawback is that they are harder to clean and require a greater flow rate of water, and so unfortunately have had little impact on the aquaculture industry.

These examples have made it clear that the problem of variability in growth can be tackled by a behavioural ecologist in a variety of ways. However, there may be economic and genetic constraints beyond our control. The slowest growing fish in a group may be genetically inferior and incapable of the same growth performance as the others. Jobling and Reinsnes (1986) found that the runts in a population of hatchery arctic charr grew better when kept in isolation (indicating that their growth was suppressed by the presence of the other fish), but never attained the average growth rate of the population.

5.1.5 How can we control maturation?

Many of the salmonids only breed once: most members of the various species of Pacific salmon (*Oncorhynchus* spp.) die after spawning, and only a small proportion of wild Atlantic salmon will survive to spawn a second time. Since most species of fish are capable of maturing at a range of ages, it is desirable for the fish farmer to defer the age of sexual maturity as long as possible, since large fish tend to be worth more per unit weight than small. The fecundity of females is strongly dependent on their size (since small females cannot carry large numbers of eggs), so that they tend to mature only when a reasonable size has been attained. However, males can mature at much smaller sizes, since sperm production is less constrained by adult size. Some mature male salmon may be only 1/100th of the weight of the females they mate with; these are often described as 'precocious' males, although this is rather a misleading term since it has overtones of developmental abnormality. The small males pay a cost in reduced survival. For instance, mature Atlantic salmon parr have up to six times the mortality rate of immature fish, even under hatchery conditions (Leyzerovich, 1973).

Even when the fish survive after maturation, early maturation is usually considered undesirable by the farmer, since it leads to a slowing of growth (as the fish puts resources into gonads rather than

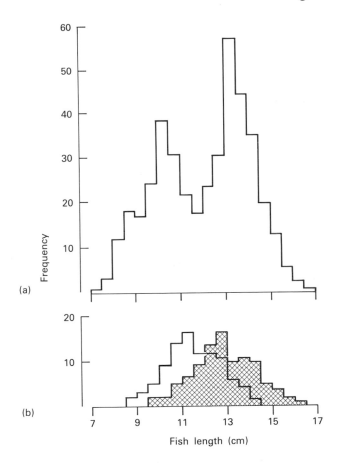

Figure 5.5 (a) A typical autumn size frequency distribution of immature Atlantic salmon parr, showing the bimodal separation into an Upper Modal Group (UMG: larger fish, that will smolt aged 1+) and Lower Modal Group (LMG: fish that will smolt a year later). (b) The size distribution of mature male parr from the same population is shown in white, while their size had they not matured is estimated in the hatched distribution. Note how most of the maturing fish would potentially have been big enough to belong to the UMG, but maturation has reduced their growth such that most fall into the LMG and so will now delay smolting. (After Saunders *et al.*, 1982.)

soma) (Figure 5.5) (Leyzerovich, 1973; Naevdal *et al.*, 1979; Saunders *et al.*, 1982). In addition, male salmon that have matured as parr (i.e. before going to sea) tend to take longer reaching the smolt stage (Saunders *et al.*, 1982; Kubo, 1983).

What causes a fish to mature? If we can understand this, it may be possible to manipulate the fish's environment or its physiological condition so that maturation is delayed until a desired time. This is where the theories of life-history strategies can play a part in managing animals. It is now clear that the triggering of sexual maturation is linked to growth rate: the faster the growth rate of the population, the greater the percentage of early-maturing fish (Alm, 1959; Naevdal *et al.*, 1979; Adams and Thorpe, 1989), and the individuals that mature first tend to be those that were growing fastest (Leyzerovich, 1973; Saunders *et al.*, 1982; Naevdal, 1983). This creates a paradox, since the fish farmer generally wants both fast growth rates *and* deferred maturity. In the wild there are population differences in the average age at maturity, which can be interpreted as adaptations to local environments (Schaffer and Elson, 1975; Scarnecchia, 1983). These differences are at least partly under genetic control (Naevdal, 1983; Sutterlin and MacLean, 1984), so that it may prove possible to selectively breed strains that exhibit the desired traits of fast growth yet late maturation. However, a quicker procedure is to restrain the growth of the fish, so that maturation is not triggered. An interesting example is the masu salmon (*Oncorhynchus masou*), which in the wild has a high incidence of early maturity in male parr, due to the fast growth rates achieved in its productive natural habitat. Hatchery fish are deliberately kept in cooler water, so reducing growth rates and hence the incidence of early maturity (Kubo, 1983; Thorpe, 1986 and personal communication). It may be possible to achieve the same results by a more selective reduction in growth rates, since the decision as to whether or not to mature in a given year may be made during a relatively short time-window. If feeding rates are much reduced during this critical period, the fish may be 'fooled' into delaying maturation for another year. Preliminary results from Atlantic salmon suggest that this approach may prove highly successful (Rowe and Thorpe, 1990; Thorpe *et al.*, 1990).

5.2   RELEASING REARED FISH

5.2.1 What is the object of releasing fish?

There are three main purposes in rearing fish for later release, which may be labelled enhancement, stocking and ranching. Since fish life-cycles usually involve a high fry and juvenile mortality, so that large numbers of eggs produce very few maturing adults, it is theoretically possible to enhance the adult population by artificially increasing juvenile survival. Survival rates in fish hatcheries commonly exceed 80% over the first year, compared with under 10% in the wild, so that releasing fish once they have passed the most vulnerable phase of the life-cycle can potentially increase harvests enormously (provided that density-dependent mortality in the later life-stages is not strong enough to offset any increases in population size). An example of this form of aquaculture is the system of State-run hatcheries on the Pacific coast of America, rearing various species of Pacific salmon.

Stocking usually refers to the practice of releasing fish in waters where there is no natural population. This can occur either because the species is not native to the area (e.g. the use of foreign species to stock lakes for recreational fishing, an example being the rainbow trout used extensively in Britain), or because its natural spawning or fry-rearing areas are no longer available to the fish, or because the water course has recently recovered from pollution that had destroyed the natural fish populations. The damming of many of Sweden's rivers in the development of hydro-electric schemes prevented adult Baltic salmon (*Salmo salar*) from reaching their traditional spawning areas. One of the conditions of allowing the dams to be built was that the power stations would also run salmon hatcheries. These maintain the salmon populations, by using the returning adults as broodstock and rearing the resulting young until they reach the (migratory) smolt stage, at which point they are released into the rivers below the dams.

The third category is an extension of the other two, in that ranching is the commercial exploitation of the biological phenomenon of homing by migratory salmonids. Since these fish will return to their natal area, they can be reared until the smolt stage, then released, and they will return as full adults at maturity. The advantage of this is in the saving of feed and caging costs during the marine phase. These are considerable, especially since the fish put on over 99% of their final weight during this time. The disadvantages are a reduction in the number of fish obtained at harvesting, due to natural and fishing mortality at sea and also a degree of imperfect homing: usually less than 10% of the smolts released will return as full adults. The problems of commercial fishing are considerable, and the system is only really viable where there is no offshore fishery. For further discussion of the principles and practices of ranching see Thorpe (1980, 1986) and Isaksson (1988).

5.2.2 What are the problems of releasing hatchery fish?

Fish released from a hatchery must switch from feeding on an artificial food source to a natural one. They have become conditioned to a simple environment which contains very little other than fish, water and excess food; they must now discriminate between food and non-food even though they may never have experienced these food types before. There is some evidence that fish can learn to exploit natural but unfamiliar food items very quickly (and even prefer them to familiar pellets) when presented with them in the laboratory (Strad-meyer and Thorpe, 1987; Ware, 1971). However, they appear to take longer in the wild, possibly due to competition from resident wild fish (Miller, 1958) and because the learning process is made more difficult by the greater variety of stimuli. Johnson and Ugedal (1986) found that hatchery-reared brown trout (*Salmo trutta*) had less full stomachs than comparable wild trout in the same river for up to a week after release. Moreover, up to 30% of the stomach contents of the hatchery fish were plant fragments (never eaten by wild fish), although they gradually learned not to eat these. Comparisons of releases made at different times of year showed that the trout learned fastest when food availability was highest, presumably because they had more opportunity to learn. This has clear implications for the optimal timing of releases, since much of the mortality of released fish occurs in the period immediately after release when the fish are adjusting to the new environment.

Stream-living fish are often reared, for reasons of practicality, at lower water velocities then they would experience in the wild. One consequence of this is that they are, in effect, physically unfit. They have less stamina than wild fish (Rimmer *et al.*, 1985) and when released are more likely to move downstream, possibly because they cannot hold station against the current (Cresswell and Williams, 1983; Shchurov *et al.*, 1986). They may also have problems feeding, since they must accelerate into the main currents to intercept food items. However, it is possible to overcome this by giving the fish exercise training before they are released. Figure 5.6 shows the results of two studies in which brown trout and Atlantic salmon were subjected to higher water flow rates prior to release. These 'trained' fish remained closer to the release site and were recaptured in greater numbers in subsequent surveys (Cresswell and Williams, 1983; Shchurov *et al.*, 1987). In addition, trained fish greatly outnumbered the controls in the faster currents of the river (the best feeding areas) (Figure 5.6b), and tended to have fuller stomachs than the controls (Shchurov *et al.*, 1987).

Probably the most serious cause of post-release mortality in reared fish is predation. This is partly because the fish, being naive, do not respond to the sight of a predator in an appropriate manner — they may even school with the predator, instead of avoiding it (Jacobsson and Jarvi, 1977). Attempts have been made to try and

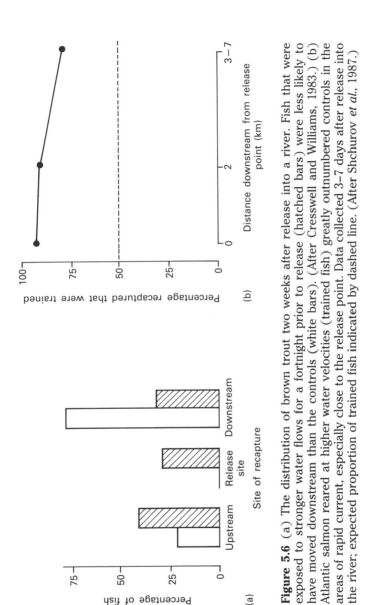

Figure 5.6 (a) The distribution of brown trout two weeks after release into a river. Fish that were exposed to stronger water flows for a fortnight prior to release (hatched bars) were less likely to have moved downstream than the controls (white bars). (After Cresswell and Williams, 1983.) (b) Atlantic salmon reared at higher water velocities (trained fish) greatly outnumbered controls in the areas of rapid current, especially close to the release point. Data collected 3–7 days after release into the river; expected proportion of trained fish indicated by dashed line. (After Shchurov *et al.*, 1987.)

train fish to avoid predators prior to release. Thompson (1966) repeatedly exposed hatchery Chinook and Coho salmon (*Oncorhynchus tshawytscha* and *O. kisutch*) fry to a model fish that emitted an electric field, to see if they would learn to avoid it. Within 13 days there was evidence of learning, although some fish persistently entered the 'danger zone' and received a shock. This learned avoidance of the model fish extended to real predators, since trained fish had half the mortality of untrained controls when groups were placed in large tanks with hungry rainbow trout (Thompson, 1966). However, Fraser (1974) had less success in a similar experiment with hatchery brook trout, using a model electrified piscivorous bird in place of a model fish. When the 'trained' trout were released into a lake populated with the same predator, they survived no better than untrained fish (16% of trained and 18% of untrained fish being found alive 10 weeks later). This may have been because in the training sessions the fish had learned to move only as far away from the approaching model bird as to take them outside its electric field (i.e. less than 0.5m), which would not be far enough to escape from a real predator.

Numerous studies have demonstrated that the risk of predation tends to be higher for solitary individuals than for members of a group (reviewed in Pulliam and Caraco, 1984; see also Chapter 3), and this is especially true for species of schooling fish (Neill and Cullen, 1974). Knowledge of the interaction between group size and predation risk can be used to reduce losses due to predation. When Lake Kinneret in Israel was stocked with silver carp (*Hypophthalmichthys molitrix*), predation by resident cichlids (*Tilapia zillii*) was greatly reduced if the silver carp were given several hours to form schools before a protecting net was removed, since the schools afforded the fish some protection (Henderson, 1980). Predators may gather when fish are released in large numbers, since this constitutes a concentrated food source (Hvidsten and Mokkelgjerd, 1987), solutions include dispersing the predators by increasing their own risk of predation or moving the release point to a less predictable location. A behavioural example of the former is the broadcasting of killer whale (*Orcinus orca*) calls at the mouths of estuaries to keep belugas (*Delphinapterus leucas*) away from the concentrations of migrating Pacific salmon (*Oncorhynchus* spp.) (Meacham and Clark, 1979), while moving the release site of hatchery Atlantic salmon smolts from an estuary (and congregations of predatory cod) to the open sea produced an estimated 29% greater survival (Gunnerod and Hvidsten, quoted in Hvidsten and Mokkelgjerd, 1987; Hansen, 1982; Larsson, 1982). An alternative approach is to try and reduce the risk to the fish by increasing the size at which they are released. Many predators will only take a limited size range of prey, so that larger fish may avoid the 'predation window' and survive better (Table 5.2; Hager and Noble, 1976; Bilton *et al.*, 1982).

Table 5.2 Relationships between sex, size at release and return rates in coho salmon smolts divided into four size categories (1=smallest) at release

Size group	Sex ratio at release (% males)	% of released males returning aged two	% of released fish returning aged three	Sex ratio in returning 3-year-olds
1	52.9	<0.1	1.55	55.2
2	51.1	1.4	2.06	49.4
3	63.1	4.7	2.15	41.0
4	69.7	10.0	2.19	33.3

Note that the larger smolts, which survive better, tend to be males. Females return aged three, whereas some males return aged two; these early-maturing males tend to have been the largest smolts. As a result the size–sex ratio relationship in the 3-year olds is opposite to that in the released smolts (After Hager and Noble, 1976)

Many commercial species are migratory, moving between fresh and salt water. Those that are born (and reproduce) in fresh water and move temporarily to sea (e.g. salmon) are called anadromous; those that conduct the reverse migration (e.g. eels) are termed catadromous (McDowall, 1988). The variability in migration strategy can, to some extent, be explained as a response to differing productivities (and hence maximal growth rates) in the two habitats: in temperate and arctic regions, marine habitats tend to be more productive than fresh-water (and so most migratory species are anadromous), while in the tropics it is the reverse (and so the majority here are catadromous) (Gross *et al.*, 1988). Thus the low productivity of the stream habitats in which juvenile salmonids are found limits the maximum size that they can attain (Bachmann, 1982). They must abandon these areas and the foraging strategy of being a sit-and-wait predator if they are to grow larger. A consequence of this is that migratory individuals tend to be much larger than residents of the same species. Jonsson (1985) found that migratory individuals of the brown trout (known as sea trout) were on average 25–75% larger than residents of the same age from the same natal area.

Attainment of a large size tends to be more important for females than for males (see earlier section on maturation). While large males may be most successful at obtaining mates by overt competition, the success of alternative mating strategies (such as sneaking) may be independent of male size, or even be highest for small males (Gross, 1985). Thus the sex ratio of migrant salmonids may be biased towards females, the males remaining behind to mature at a much smaller size (Myers, 1984; Jonsson, 1985). From an economic point of view, male salmonids that have been allowed to mature in the freshwater hatchery are then less likely to migrate to sea when released, and in fact may tolerate seawater (Dellefors and Faremo, 1988), and so will not contribute to the adult harvest (Hansen *et al.*, 1989). It is therefore desirable to delay the timing of maturation so that males migrate rather than mature in freshwater; this can be achieved by slowing down the growth rate at critical periods (see section on maturation).

Migratory salmonids return to their natal river to spawn. They recognize this by its characteristic 'smell', which they become imprinted upon during a critical period earlier in life. For most species, this critical period is shortly before or during the first migration to sea (Hasler and Scholtz, 1983). By a process of sequential imprinting the fish will learn the smell of its home stream, then home river, and so on until it has reached the sea. Ranched fish can therefore be moved between sites during the freshwater period, as long as they are kept at the site to which they are required to home during the period prior to release (Thorpe, 1986). However, if the fish are moved at the smolt stage to a release site with few predators (see above), it is important to give them the correct olfactory experience of this stage of the 'migration'. Moving the fish in a floating cage enables them to 'swim' the route and so receive the correct imprinting cues, which results in a higher return rate than if the fish are transported by road tankers (Larsson, 1982). Salmon can also be made to imprint onto any specific chemical that they are exposed to during the critical period; they will then return to a point in their home river system where this is released. In the case of fish farms, they will naturally return to the outflow of the farm, since this is the source of the 'correct' smell of fishfood plus river water that they became imprinted upon. Sockeye salmon (*Oncorhynchus nerka*), which undertake an early migration from streams to lakes after first beginning to feed, imprint on their natal site during the incubation stage. The eggs can therefore be collected from streams after spawning and incubated in a hatchery until the fish have reached the feeding stage. These are then moved to a suitable release point, and the returning adults will home on the hatchery (Brannon, 1982, quoted in Thorpe, 1988).

In conclusion, I hope that this chapter has made it clear that an understanding of the behaviour and ecology of commercial fish species is vital if we are to manage them successfully. The aquaculture industry needs to be able to maximize the growth rate of large numbers of fish kept at high densities. We have seen that this can create a multitude of behavioural problems, which are not necessarily insolvable provided that we understand the causes. There is also a need to control the developmental pathways of the fish by means of manipulating their life-history strategies, and increase their survival if they are released by enhancing their anti-predator responses. A behavioural approach will not solve everything, but there is clearly a need for ethologists and behavioural ecologists to persuade fish farmers to stop treating fish as machines!

REFERENCES

Abbott, R.R. (1972) Induced aggregation of pond-reared rainbow trout (*Salmo gairdneri*) through acoustic conditioning. *Trans. Am. Fish. Soc.*, **101**, 35–43.

Abbott, J.C. and Dill, L.M. (1985) Patterns of aggressive attack in juvenile steelhead trout (*Salmo gairdneri*). *Can. J. Fish. Aquat. Sci.*, **42**, 1702–6.

Adams, C.E. and Thorpe, J.E. (1989) Photoperiod and temperature effects on early development and reproductive investment in Atlantic salmon (*Salmo salar* L.). *Aquaculture*, **79**, 403–9.

Alm, G. (1959) Connection between maturity, size and age in fishes. *Rep. Inst. Freshwat. Res. Drottningholm*, **40**, 5–145.

Bachmann, R.A. (1982) A growth model for drift-feeding salmonids: a selective pressure for migration, in *Proceedings of Salmon and Trout Migratory Behavior Symposium* (eds E.L. Brannon and E.O. Salo), pp. 128–35, University of Washington, Seattle, USA.

Bilton, H.T., Alderdice, D.F. and Schnute, J.T. (1982) Influence of time and size at release of juvenile Coho salmon (*Oncorhynchus kisutch*) on returns at maturity. *Can. J. Fish. Aquat. Sci.*, **39**, 429–47.

Brannon, E.L. (1982) Orientation mechanisms of homing salmonids, in *Proceedings of the Salmon and Trout Migratory Behavior Symposium* (eds E.L. Brannon and E.O. Salo), pp. 219–27, University of Washington, Seattle, U.S.A.

Clarke, L.A. and Sutterlin, A.M. (1985) Associative learning, short-term memory, and colour preference during first feeding by juvenile Atlantic salmon. *Can. J. Zool.*, **63**, 9–14.

Cowey, C.B., Mackie, A.M. and Bell, J.G. (eds) (1985) *Nutrition and Feeding in Fish*, Academic Press, London.

Cresswell, R.C. and Williams, R. (1983) Post-stocking movements and recapture of hatchery-reared trout released into flowing waters —

effect of prior acclimation to flow. *J. Fish Biol.*, **23**, 265–76.

Davies, N.B. and Houston, A.I. (1984) Territory economics, in *Behavioural Ecology: an Evolutionary Approach*, 2nd edn (eds J.R. Krebs and N.B. Davies), pp. 148–69. Blackwell Scientific Publications, Oxford.

Dellefors, C. and Faremo, U. (1988) Early sexual maturation in males of wild sea trout, *Salmo trutta* L., inhibits smoltification, *J. Fish Biol.*, **33**, 741–9.

Dunbrack, R.L. and Dill, L.M. (1983) A model of size dependent surface feeding in a stream dwelling salmonid. *Env. Biol. Fish.*, **8**, 203–16.

Elliott, J.M. (1982) The effects of temperature and ration size on the growth and energetics of salmonids in captivity. *Comp. Biochem. Physiol.*, **73B**, 81–91.

Fausch, K.D. (1984) Profitable stream positions for salmonids: relating specific growth rate to net energy gain. *Can. J. Zool.*, **62**, 441–51.

Ferguson, M.M. and Noakes, D.L.G. (1982) Genetics of social behaviour in charrs (*Salvelinus* species). *Anim. Behav.*, **30**, 128–34.

Ferno, A. and Holm, M. (1986) Aggression and growth of Atlantic salmon parr. I. Different stocking densities and size groups. *FiskDir. Skr. Ser. HavUnders.*, **18**, 113–22.

Fraser, J.M. (1974) An attempt to train hatchery-reared brook trout to avoid predation by the common loon. *Trans. Am. Fish. Soc.*, **103**, 815–18.

Ginetz, R.M. and Larkin, P.A. (1973) Choice of colors of food items by rainbow trout (*Salmo gairdneri*). *J. Fish. Res. Board Can.*, **30**, 229–34.

Gowan, R.J. and Bradbury, N.B. (1987) The ecological impact of salmonid farming in coastal waters: a review. *Oceanogr. Mar. Biol. Ann. Rev.*, **25**, 563–75.

Gross, M.R. (1985) Disruptive selection for alternative life histories in salmon. *Nature*, **313**, 47–8.

Gross, M.R., Coleman, R.M. and McDowall, R.M. (1988) Aquatic productivity and the evolution of diadromous fish migration. *Science*, **239**, 1291–3.

Hager, R.C. and Noble, R.E. (1976) Relation of size at release of hatchery-reared Coho salmon to age, size and sex composition of returning adults. *Prog. Fish-Cult.*, **38**, 144–7.

Halver, J.E. and Tiews, K. (eds) (1979) *Finfish Nutrition and Fishfeed Technology* (2 vols), Heinemann, Berlin.

Hansen, L.P. (1982) Salmon ranching in Norway, in *Sea Ranching of Atlantic Salmon* (eds C. Eriksson, M.P. Ferranti and P.O. Larsson), pp. 95–108, COST 46/4 Workshop, EEC, Brussels.

Hansen, L.P., Jonsson, B., Morgan, R.I.G. and Thorpe, J.E. (1989) Influence of parr maturity on emigration of smolting Atlantic salmon (*Salmo salar*). *Can. J. Fish. Aquat. Sci.*, **46**, 410–15.

Hasler, A.D. and Scholtz, A.T. (1983) *Olfactory Imprinting and Homing in Salmon*, Springer-Verlag, Berlin.

Henderson, H.F. (1980) Behavioral adjustment of fishes to release into a new habitat, in *Fish Behavior and its Use in the Capture and Culture of Fishes* (eds J.E. Bardach, J.J. Magnusson, R.C. May and J.M. Reinhart), pp. 331–44, ICLARM Conference Proceedings Vol. 5, ICLARM, Manilla, Philippines.

Hvidsten, N.A. and Mokkelgjerd, P.I. (1987) Predation on Salmon smolts, *Salmo salar* L., in the estuary of the River Surna, Norway. *J. Fish Biol.*, **30**, 273–80.

Isaksson, A. (1988) Salmon ranching: a world view. *Aquaculture*, **75**, 1–33.

Jacobsson, S. and Jarvi, T. (1977) (Anti-predator behaviour of 2-year old hatchery reared Atlantic salmon *Salmo salar* and a description of the predatory behaviour of burbot *Lota lota*). *Zool. Revy*, *38*, 57–70. (In Swedish.)

Jakobsen, P.J., Johnsen, G.H. and Holm, J. Chr. (1987) Increased growth rate in Atlantic salmon parr (*Salmo salar*) by using a two-coloured diet. *Can. J. Fish. Aquat. Sci.*, **44**, 1079–82.

Jobling, M. (1983) Effect of feeding frequency on food intake and growth of Arctic charr, *Salvelinus alpinus* L. *J. Fish Biol.*, **23**, 177–85.

Jobling, M. (1985). Physiological and social constraints on growth of fish with special reference to Arctic charr, *Salvelinus alpinus* L. *Aquaculture*, **44**, 83–90.

Jobling, M. and Reinsnes, T.G. (1986) Physiological and social constraints on growth of Arctic charr, *Salvelinus alpinus* L.: an investigation of factors leading to stunting. *J. Fish Biol.*, **28**, 379–84.

Jobling, M. and Reinsnes, T.G. (1987). Effect of sorting on size-frequency distributions and growth of Arctic charr, *Salvelinus alpinus* L. *Aquaculture*, **60**, 27–31.

Johnsen, B.O. and Ugedal, O. (1986) Feeding by hatchery-reared and wild brown trout, *Salmo trutta* L., in a Norwegian stream. *Aquacult. Fish. Mgmt.*, **17**, 281–7.

Jonsson, B. (1985) Life history patterns of freshwater resident and sea-run migrant brown trout in Norway. *Trans. Am. Fish. Soc.*, **114**, 182–94.

Knights, B. (1983) Food particle-size preferences and feeding behaviour in warmwater aquaculture of European eel, *Anguilla anguilla* (L.). *Aquaculture*, **30**, 173–90.

Koebele, B.P. (1985) Growth and the size hierarchy effect: an experimental assessment of three proposed mechanisms: activity differences, disproportional food acquisition, physiological stress. *Env. Biol. Fish.*, **12**, 181–8.

Konikoff, M. and Lewis, W.M. (1974) Variation in weight of cage-reared channel catfish. *Prog. Fish-Cult.*, **36**, 138–44.

Krebs, J.R., Stephens, D.W. and Sutherland, W.J. (1983) Perspectives in optimal foraging, in *Perspectives in Ornithology* (eds G.A. Clark and A.H. Brush), pp. 165–216, Cambridge University Press, New York.

Kubo, T. (1983) Growth-rate and smolting-rate of anadromous 'Masu' salmon (*Oncorhynchus masou*) juveniles under artificial conditions. *Scient. Rep. Hokk. Salmon Hatch.*, **37**, 23–39.

Landless, P.J. (1976) Demand-feeding behaviour of rainbow trout. *Aquaculture*, **7**, 11–25.

Larson, P.O. (1982) Salmon ranching in Sweden, in *Sea Ranching of Atlantic Salmon* (eds C. Eriksson, M.P. Ferranti and P.O. Larsson), pp. 127–37, COST 46/4 Workshop, EEC, Brussels.

Leyzerovich, Kh. A. (1973) Dwarf males in hatchery propagation of the Atlantic salmon (*Salmo salar* L.). *J. Ichthyol.*, **13**, 382–92.

McDowall, R.M. (1988) *Diadromy in Fishes*, Croom Helm, London.

Meacham, C.P. and Clark, J.H. (1979) Management to increase anadromous salmon production, in *Predator-Prey Systems in Fisheries Management* (eds R.H. Stroud and H. Clepper), pp. 377–86, Sport Fishing Institute, Washington DC.

Metcalfe, N.B. (1986) Intraspecific variation in competitive ability and food intake in salmonids: consequences for energy budgets and growth rates. *J. Fish Biol.*, **28**, 525–31.

Metcalfe, N.B. (1989) Differential response to a competitor by Atlantic salmon adopting alternative life-history strategies. *Proc. R. Soc. Lond.*, **B236**, 21–7.

Metcalfe, N.B., Huntingford, F.A., Graham, W.D. and Thorpe, J.E. (1989) Early social status and the development of life-history strategies in Atlantic salmon. *Proc. R. Soc. Lond.*, **B236**, 7–19.

Metcalfe, N.B., Huntingford, F.A. and Thorpe, J.E. (1986) Seasonal changes in feeding motivation of juvenile Atlantic salmon (*Salmo salar*). *Can. J. Zool.*, **64**, 2439–46.

Metcalfe, N.B., Huntingford, F.A. and Thorpe, J.E. (1988) Feeding intensity, growth rates, and the establishment of life-history patterns in juvenile Atlantic salmon. *J. Anim. Ecol.*, **57**, 463–74.

Milinski, M. (1984) A predator's costs of overcoming the confusion-effect of swarming prey. *Anim. Behav.*, **32**, 1157–62.

Miller, R.B. (1958) The role of competition in the mortality of hatchery trout. *J. Fish. Res. Bd. Canada*, **15**, 27–45.

Mork, O.I. (1982) Growth of three salmonid species in mono and double culture (*Salmo salar* L., *S. trutta* L. and *S. gairdneri* Rich.). *Aquaculture*, **27** 141–7.

Myers, R.A. (1984) Demographic consequences of precocious maturation of Atlantic salmon (*Salmo salar*). *Can. J. Fish. Aquat. Sci.*, **41**, 1349–53.

Naevdal, G. (1983) Genetic factors in connection with age at maturation. *Aquaculture*, **33**, 97–106.

Naevdal, G., Holm, M., Leroy, R. and Moller, D. (1979). Individual growth rate and age at sexual maturity in rainbow trout. *FiskDir. Skr. Ser. HavUnders.*, **17**, 1–10.

Neill, S.R. St J. and Cullen, J.M. (1974) Experiments on whether schooling by their prey affects the hunting behaviour in cephalopods and fish predators. *J. Zool. (Lond.)*, **172**, 549–69.

Noakes, D.L.G. and Grant, J.W.A. (1986) Behavioural ecology and production of riverine fishes. *Pol. Arch. Hydrobiol.*, **33**, 249–62.

Peters, G., Delventhal, H. and Klinger, H. (1980) Physiological and morphological effects of social stress in the eel (*Anguilla anguilla* L.). *Arch. FischWiss.*, **30**, 157–80.

Phillips, M.J. (1982) The attraction of free-ranging rainbow trout to a feeding station. Unpubl. Ph.D. Thesis, University of Stirling, U.K.

Pickering, A.D., Griffiths, R. and Pottinger, T.G. (1987) A comparison of the effects of overhead cover on the growth, survival and haematology of juvenile Atlantic salmon, *Salmo salar* L., brown trout *Salmo trutta* L., and rainbow trout *Salmo gairdneri* Richardson. *Aquaculture*, **66**, 109–24.

Pitcher, T.J. and Hart, P.J.B. (1982) *Fisheries Ecology*, Croom Helm, London.

Pulliam, H.R. and Caraco, T. (1984) Living in groups: is there an optimal group size?, in *Behavioural Ecology: an Evolutionary Approach*, 2nd edn (eds J.R. Krebs and N.B. Davies), pp. 122–47 Blackwell Scientific Publications, Oxford.

Refstie, T. and Kittelsen, A. (1976) Effect of density on growth and survival of artificially reared Atlantic salmon. *Aquaculture*, **8**, 319–26.

Rimmer, D.M. and Power, G. (1978) Feeding response of Atlantic salmon (*Salmo salar*) alevins in flowing and still water. *J. Fish. Res. Board Can.*, **35**, 329–32.

Rimmer, D.M., Saunders, R.L. and Paim, U. (1985) Effects of temperature and season on the position holding performance of juvenile Atlantic salmon (*Salmo salar*). *Can. J. Zool.*, **63**, 92–6.

Ringler, N.H. (1979) Selective predation by drift-feeding brown trout (*Salmo trutta*). *J. Fish. Res. Board Can.*, **36**, 392–403.

Rowe, D.K. and Thorpe, J.E. (1990) Suppression of maturation in male

Atlantic salmon (*Salmo salar* L.) parr by reduction in feeding and growth during spring months. *Aquaculture*, in press.

Saclauso, C.A. (1985). Interaction of growth with social behaviour in *Tilapia zillii* raised in three temperatures. *J. Fish Biol.*, **26**, 331–7.

Saunders, R.L., Henderson, E.B. and Glebe, B.D. (1982) Precocious sexual maturation and smoltification in male Atlantic salmon (*Salmo salar*). *Aquaculture*, **28**, 211–29.

Scarnecchia, D.L. (1983) Age at sexual maturity in Icelandic stocks of Atlantic salmon (*Salmo salar*). *Can. J. Fish. Aquat. Sci.*, **40**, 1456–68.

Schaffer, W.M. and Elson, P.F. (1975) The adaptive significance of variations in life history among local populations of Atlantic salmon in North America. *Ecology*, **56**, 577–90.

Shchurov, I.L., Smirnov, Yu.A. and Shustov, Yu.A. (1986) Features of adaptation of hatchery young of Atlantic salmon, *Salmo salar*, to riverine conditions after a conditioning period before release. I. Possibility of conditioning of the young under hatchery conditions. *J. Ichthyol.*, **26**, 26–30.

Shchurov, I.L., Smirnov, Yu.A. and Shustov, Yu.A. (1987) Features of adaptation of hatchery young of Atlantic salmon, *Salmo salar*, to river conditions. II. Behavior and feeding conditioning in the river. *J. Ichthyol.*, **27**, 162–6.

Slaney, P.A. and Northcote, T.G. (1974) Effects of prey abundance on density and territorial behavior of young rainbow trout (*Salmo gairdneri*) in laboratory stream channels. *J. Fish. Res. Board Can.*, **31**, 1201–9.

Smith, J.J. and Li, H.W. (1983) Energetic factors influencing foraging tactics of juvenile steelhead trout, *Salmo gairdneri*, in *Predators and Prey in Fishes* (ed. D.L.G. Noakes), pp. 173–80, Dr W. Junk, The Hague.

Stradmeyer, L. and Thorpe, J.E. (1987) The responses of hatchery-reared Atlantic salmon, *Salmo salar* L., parr to pelleted and wild prey. *Aquacult. Fish. Manage.*, **18**, 51–61.

Stradmeyer, L., Metcalfe, N.B. and Thorpe, J.E. (1988) Effect of food pellet shape and texture on the feeding response of juvenile Atlantic salmon. *Aquaculture*, **73**, 217–28.

Sutterlin, A.M. and MacLean, D. (1984) Age at first maturity and the early expression of oocyte recruitment processes in two forms of Atlantic salmon (*Salmo salar*) and their hybrids. *Can. J. Fish. Aquat. Sci.*, **41**, 1139–49.

Taylor, E.B. (1988) Adaptive variation in rheotactic and agonistic behaviour in newly-emerged fry of chinook salmon, *Oncorhynchus tshawytscha*, from ocean- and stream-type populations. *Can. J. Fish.*

Aquat. Sci., **45**, 237-43.

Taylor, E.B. and Larkin, P.A. (1986) Current response and agonistic behavior in newly emerged fry of chinook salmon, *Oncorhynchus tshawytscha*, from stream- and ocean-type populations. *Can. J. Fish. Aquat. Sci.*, **43**, 565–73.

Thompson, R.B. (1966) Effects of predator avoidance conditioning on the post-release survival rate of artificially propagated salmon. Unpubl. Ph.D. Thesis, University of Washington, Seattle.

Thorpe, J.E. (1977) Bimodal distribution of length of juvenile Atlantic salmon under artificial rearing conditions. *J. Fish Biol.*, **11**, 175–84.

Thorpe, J.E. (ed.) (1980) *Salmon Ranching*, Academic Press, London.

Thorpe, J.E. (1986) Some biological problems in ranching salmonids. *Rep. Inst. Freshw. Res. Drottningholm*, **63**, 91–104.

Thorpe, J.E. (1987) Environmental regulation of growth patterns in juvenile Atlantic salmon, in *Age and Growth of Fish* (eds R.C. Summerfelt and G.E. Hall), pp. 463–74, Iowa State University Press, Ames, Iowa, USA.

Thorpe, J.E. (1988) Salmon migration. *Sci. Prog. (Oxf.)*, **72**, 345–70.

Thorpe, J.E. and Wankowski, J.W.J. (1979) Feed presentation and food particle size for juvenile Atlantic salmon, *Salmo salar* L., in *Finfish Nutrition and Fishfeed Technology* (eds J.E. Halver and K. Tiews), pp. 501–13, Heinemann, Berlin.

Thorpe, J.E., Talbot, C., Miles, M.S. and Keay, D.S. (1990) Control of maturation in cultured Atlantic salmon, *Salmo salar* in pumped seawater tanks, by restricting food intake. *Aquaculture*, in press.

Tipping, J.M., Rathjvon, R.L. and Moore, S.T. (1986) Use of demand feeders on large steelhead rearing ponds. *Prog. Fish-Cult.*, **48**, 303–4.

Wankowski, J.W.J. (1979) Morphological limitations, prey size selectivity, and growth response of juvenile Atlantic salmon, *Salmo salar*. *J. Fish Biol.*, **14**, 89–100.

Wankowski, J.W.J. (1981) Behavioural aspects of predation by juvenile salmon (*Salmo salar* L.) on particulate prey. *Anim. Behav.*, **29**, 557–71.

Ware, D.M. (1971) Predation by rainbow trout (*Salmo gairdneri*): the effect of experience. *J. Fish. Res. Board Can.*, **28**, 1847–52.

Werner, E.E. (1974) The fish size, prey size, handling time relation in several sunfishes and some implications. *J. Fish. Res. Board Can.*, **31**, 1531–6.

Werner, E.E. and Hall, D.J. (1974) Optimal foraging and the size selection of prey by the bluegill sunfish *Lepomis macrochirus. Ecology*, **55**, 1042–52.

Yamagishi, H. (1962) Growth relation in some small experimental populations of rainbow trout fry, *Salmo gairdneri* Richardson with special reference to social relations among individuals, *Jpn. J. Ecol.,* **12**, 43–53.

6

Agriculture

DAVID G.M. WOOD-GUSH

One species of common domesticated animal and one of the most common of agricultural species, the fowl, has had a long history of involvement in animal behaviour studies (Wood-Gush, 1955) but until relatively recently the behaviour of domesticated animals was generally neglected by ethologists. It was generally, but erroneously, assumed that their behaviour was 'degenerate', although as Price (1984) and others have shown, the changes in behaviour, that domestication appears to have wrought, are quantitative rather than qualitative. Indeed in some respects their behaviour as a result of domestication has become more interesting with regard to both pure and applied behavioural science. This chapter illustrates how the application of ethology may serve agriculture.

6.1 ANIMAL PRODUCTION

6.1.1 Social behaviour, productivity and use of resources

This section is limited to cattle, sheep, pigs and domestic fowls. Under feral conditions with ecologically rich environments and in which human restraint is minimal or absent, these species show social structures different from those found in agriculture. The wild cattle of Chillingham occupy a parkland of 150ha and under these conditions the females and young stock roam freely but the bulls are in three groups with different home ranges. Young bulls may change from one bull group to another or join the main herd (Hall, 1979). The Soay sheep on the island of St Kilda, on the west coast of Scotland, form separate ewe and male groups which only mix during the rut (Grubb and Jewell, 1966). Studies on a small group of pigs in a semi-natural enclosure near Edinburgh showed that they tended to split themselves into very small sub-groups of closely related females with the boar being largely independent (Stolba and Wood-Gush, 1984). However, such sub-groups shared a nest at night if they were well acquainted. This tendency to live in small groups was also reported by Krosniunas (1984) who studied feral pigs on Santa Cruz Island off the coast of California, where large temporary aggregations might sometimes form at sources rich in food. Feral chickens studied by McBride *et al.* (1969) on an island off the coast of Queensland have a territorial social system, each territory being occupied by a dominant cock, 4–6 hens, pullets and several subordinate males. Within such a group a hierarchy is found. Some of the subordinate males may show varying degrees of territoriality, some defending a very small territory into which the dominant male may intrude. Normally the territories provide the group with all their needs but during the breeding season the hens with broods become solitary and occupy distinct home ranges.

Under many agricultural conditions in which animals are kept in very large monocaste populations, undergoing mixing of strange animals, the social structures are different. There is intense competition in crowded conditions for several resources. Overt aggression is frequent and social rank or competitiveness may be expected to affect productivity (see Chapters 3 and 5). Many workers therefore have concentrated on studying the dominance-subordinate relationships of the animals and looked for dominance hierarchies. While linear hierarchies in large groups are unlikely (Appleby, 1983) it is often possible to classify individuals as high ranking, middle and low ranking. In sheep, however, such social rankings are less obvious although some sheep are more competitive than others in intensive conditions (Syme and Syme, 1979). While some studies have shown social rank to be positively related to productivity there have been exceptions, but these results may have been due to faults in the measurement of dominance, e.g. assuming dominance in one context to be equivalent to dominance in another context (Syme, 1974). Furthermore the distribution of resources may significantly affect the findings as suggested by the following two cases.

In one study on the effect of social organization on productivity in laying hens, McBride (1962) kept 510 hens under three different conditions; individually caged, in a deep litter house with access to outside runs and a similar house with no outside access (the 'intensive' house). The birds were matched for genotype over the three environments, a pair of full sisters being present in each one. In the 'intensively' housed birds there were a significantly larger number of birds producing less than the average number of eggs compared with the caged birds. The curve for egg production was normally distributed in the caged birds but significantly skewed in the intensively housed birds. Egg weights were sampled in the last month of the year and a similar result was found. In a later paper McBride (1964) described scoring birds from these two environments for their ability to win fights over the individuals in six panels of birds. Scoring the birds on a scale from one to five he called this their aggressiveness score which he assumed to be highly correlated with their peck order status. Then from the record of the number of eggs and the birds' sample of egg weights he calculated the egg mass produced by each bird. When the relationship between egg mass and the so-called aggressiveness score was examined, it was found that in the caged birds the relationship was non-significant but in the intensively housed birds the regression of egg mass on aggressiveness score was significant ($P < 0.001$). While this study has a number of imperfections, such as the use of a single competitive test to assess relative status, the equating of aggression with fighting skill, the use of eggs from a single month to assess egg weight and hence egg mass, the results are highly suggestive that low social status affects productivity when the birds are housed on the floor. In the second study the effect of group housing on dominance and productivity was also examined by Hughes (1977). Social dominance was measured using three scores. No relationship was found and in fact the level of aggression was very low in many cages. The differences between these two findings could be due to a number of factors, one of which could be the distribution of resources and their availability to all the hens.

McBride and his co-workers (McBride *et al.*, 1964, 1965) investigated the effects of social rank on the growth rate of pigs up to 16 weeks of age. Soon after birth each piglet in a litter tends to form a preference for a particular teat on the sow and this is known as the teat order. The anterior teats are more productive than the posterior teats. McBride and his co-workers observed 38 litters from birth to weaning at 8 weeks of age and while at present most piglets are weaned at 3 weeks of age, the two studies are of interest in showing the effect of social behaviour on growth. At weaning the piglets were divided into groups of 6–10 often composed of animals from different litters. Analysis showed that piglets in the top half of the social order at 8 weeks of age were 8kg heavier than those in the bottom half. By 16 weeks this difference was of the order of 11kg. At both 8 and 16 weeks of age about 15% of the variation in body weights could be attributed to social rank. McBride *et al.* (1965) summarized the factors affecting growth in piglets up to 16 weeks of age in Figure 6.1.

In dairy cows there appears to be no clear-cut relationship between milk production and social status, for several studies have reported contradictory results (Schein and Fohrman, 1955; Beilharz *et al.*, 1966; Dickson *et al.* (1967); Barton *et al.*, 1974; Arave and Albright, 1976;

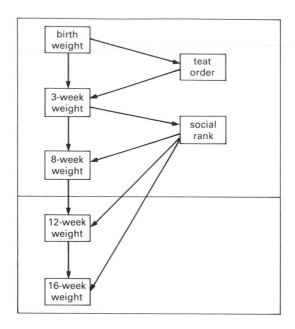

Figure 6.1 Paths of influence affecting variation in growth within litters in the pig. (From McBride *et al.*, 1965.)

Collis, 1976; Konggaard and Kohn, 1976; Soffie *et al.*, 1976; Friend and Polan, 1978; Sambraus *et al.*, 1978/9; Sambraus *et al.*, 1979; Anderson *et al.*, 1984). However, feeding methods, bedding arrangements and overall space have differed between the studies. The most clear-cut effect of social factors on production in dairy cattle is shown by the work of Krohn and Konggaard (1979). They observed three herds, in which some heifers calving for the first time were kept as a group and others mixed with older cows calving at the same time. In all three cases the heifers kept as a group spent more time feeding spread out over more sessions of eating. In one herd in which food consumption could be measured, the heifers in the heifer groups took in more roughage than the heifers in the mixed age group and, over the three herds, the heifers in the heifer group produced 5–10% more milk than the other heifers.

Summarizing the findings from the three species discussed it seems plausible to suggest that where vital resources are not carefully distributed severe competition can ensue. This is probably so in the pens of growing pigs (e.g. Ewbank and Meese, 1971) and so McBride's results are not surprising. In many dairy herds considered and in battery cages with three birds to a cage, resources are probably available to all the animals, competition may be absent or mild, and dominance therefore will have had little effect on productivity. While it must be borne in mind that, in some instances, low status may be held by animals that are inherently poor producers and high status by good producers, studies suggest that the distribution of resources and hence the design of housing systems is very important. Nevertheless, while this may seem like a truism, the designers of animal housing have usually had other priorities in mind such as the economics of construction, disease control, temperature control, cleansing and human convenience. Provision has been aimed at the average animal, and social interactions have only been considered when severe problems such as feather-pecking in fowls or tail biting in pigs have occurred. Only very recently, with the advent of cheaper electronic equipment, has individual feeding become more frequent. Even here, however, behavioural problems can arise from a poor layout of the equipment and food is but one resource.

6.1.2 Mating behaviour

An interesting example of how a management practice can have detrimental effects on the sexual behaviour of farm animals was shown by the work of Hemsworth *et al.* (1977). In Australia many pig breeders housed their young breeding boars in individual pens to enable the breeders accurately to assess the growth rate and feed conversion abilities of the animals. However, the sexual performance of such boars was often below expectation. Hemsworth *et al.* reared three groups of six boars from 20 days of age, a common weaning age in industry, to 7 months under one of the following social conditions: a mixed sex group in which the gilts were sexually immature; an all-male group; or in social isolation, without physical and visual contact with other pigs. The sexual behaviour of these males was then studied from 7 to 13 months of age during which period they were all individually housed in identical pens. Starting at 30 weeks of age these boars were given 20 minute mating tests at 2 week intervals for 6 months, using spayed gilts treated with progesterone and oestradiol benzoate. The physical development of the boars was the same for all groups, but the males reared in early isolation differed from the males in the other two groups in that their courtship contained fewer courting behaviour activities per unit time than the other groups, which showed vigorous courtship behaviour and which did not differ from each other. These differences were mainly due to significant differences ($P < 0.05$) in the total number of mounts, nosing the sides of the gilt, chants and anogenital sniffs. In addition the total number of copulations and total time spent ejaculating over the 12 mating tests were significantly less ($P<0.01$) for the boars reared in isolation. Contrary to expectation, however, there were no differences in the mating dexterity of the males from the three groups. However three of the boars reared in isolation produced semen with high percentages of abnormal sperm.

A further study (Hemsworth *et al.*, 1978) revealed that lack of physical contact with other pigs accounted for 70% of the impairment in the copulatory performance of boars reared in the absence of visual and physical contact with other pigs. Furthermore, boars kept in visual and physical isolation from the age of 12 weeks showed no significant differences in copulatory performance from group-reared males, although they tend to have lower scores, which the authors suggest might be significant when tested in a large population. (For the use of artificial pheromones in pig matings see Chapter 4.)

Non-assortative mating may also reduce fertility and possibly inter-rupt breeding programmes in which Animal Productionists generally assume that all matings will be random (see Chapter 3). In an early study Wood-Gush (1954) reported an experiment with Brown Leghorn fowls in which hens had solicited the cocks at significantly different rates. However, the basis of the hens' preferences was not discovered, for while the males had similar plumage they had differed quantita-tively in their courtship behaviour and in some physical characteristics such as size. In a further experiment (Lill and Wood-Gush, 1965), nine cocks, from the same breeding line as used previously, were tested twice individually in two pens, each with nine sexually experienced hens, for 15 minutes. Again, the hens solicited the cocks at signifi-cantly different rates. Examination of the courtship of the males indicated that two particular behaviour patterns, the Waltz and the Rear Approach, were frequently performed before a hen voluntarily gave a sexual crouch. In the Waltz (see Figure 6.2) the cock drops one wing, partially retracts the next and shuffles round the hen with small side steps. In the Rear Approach, the cock approaches the hen from behind with a very high stepping gait. Although important, the relationship between these two displays and the solicitation by the female was not a simple one and the authors pointed out that the females often apparently discrminiated on visual cues based on small, subtle morphological differences. In another experiment involving Brown Leghorn, White Leghorn and Broiler fowls which had been reared with their own breed only, the Brown Leghorn and White Leghorn hens significantly preferred cocks from their own breed. Broiler hens, on the other hand, solicited Brown Leghorn cocks more than the cocks of their own breed, making it difficult to implicate an

Figure 6.2 A Brown Leghorn Cock waltzing to a female. (From Wood-Gush, 1971.)

imprinting process as the main cause for the preferences found in the experiment (Lill and Wood-Gush, 1965).

Male preferences were also examined by allowing each cock a choice between two hens presented simultaneously. The experimental set-up is shown in Figure 6.3. In each of two cages was a hen of a different breed and the cocks were released singly into the observation pen, at a point equidistant from the two hens. The areas A and B surrounding the hen cages were of equal area and the time spent in each cage area by the cock and his courtship behaviour were scored. Each cock was tested twice in three separate discriminations: Brown Leghorn/Broiler, Brown Leghorn/White Leghorn and Broiler/White Leghorn. In the second test, carried out 7 days later, the positions of the hens were reversed. Six cocks of each breed which had been reared with their own breed were used, as well as five Brown Leghorn and four Broiler males which had been reared in a mixed flock comprising birds of the three breeds. Brown and White Leghorn cocks with early experience only of their own breed preferred hens of their own breed, but in the Broilers with experience only of their breed the preference was less pronounced. However, in cocks with early experience of the

Figure 6.3 A Brown Leghorn Cock avoids a White Leghorn hen and waltzes to a Brown Leghorn Hen in a choice experiment. (From Wood-Gush, 1971.)

other breeds these preferences were not significant. These finding indicated that if cross breeding is to be performed involving strains or breeds differing in plumage or behaviour then mixed rearing would be advisable to avoid some infertility.

Mating preferences have also been reported in the ram (Tilbrook and Lindsay, 1987). Key and MacIver (1980) observed the sexual choices of Clun Forest and Welsh Mountain rams. The Clun Forest breed is a large fine woolled sheep from Welsh–English borders, while the Welsh Mountain sheep is much smaller with a thick fleece of coarse wool. Observations were made on eight rams, four of which had been fostered from birth onto ewes of the other breed and four of which had been reared by their biological dams. They were tested with ewes of the same breeds and rearing conditions, except that half of the ewes had been reared in a distant pasture and half had been reared either in the same field as the rams, or in an adjacent field until weaning, when the rams had been removed to a field isolated from the ewe lambs. For the preference tests the rams were released amongst the ewes for 1 to 2 hours a day. As the ewes were still somewhat immature, preferences were judged by courtship rather than copulation. Two of the rams showed no interest in the ewes but the other six gave interesting results. They consisted of two Clun Forest rams reared by their biological mothers, two Clun Forest rams reared by Welsh Mountain ewes, one Welsh Mountain ram reared by its biological mother and one Welsh Mountain ram reared by a Clun Forest ewe. Four showed a significant preference for ewes of the same breed as the ewe that had reared them, whether they had been cross fostered or not. However, most interesting was the finding that they preferred ewes that had been reared in a distant, as opposed to the same or adjacent, pasture as themselves. Although these numbers are very small, it is interesting to note that Geist (1971) reported that Bighorn and Stones rams in North America migrate during the mating season to specific rutting grounds away from their normal home ranges, although ewes may be available in those areas.

Attempts were made by Tilbrook and Lindsay (1987) to investigate the attractiveness of Merino ewes. Using two groups of seven ovariectomized ewes brought into oestrus by exogenous hormones, they ran tests to see if different rams would rank these ewes similarly for attractiveness. Ten Merino males with sexual experience were used for the tests in which six ewes were introduced to each ram in a small test pen. When the ram had spent 5 minutes courting a ewe she was deemed to be the most attractive and removed. He then moved onto another which, after 5 minutes of his attention, was also removed. In this way the attractiveness of the ewes was assessed. The ten rams were found to rank the ewes similarly and when the tests were repeated the relative ranks of the females remained the same. As the ewe plays an active role in the courtship the authors compared the attractiveness of the ewes when they were free or tethered in such a way to prohibit some of their normal courtship. Tethering was not found to affect the relative rankings of the ewes and the authors concluded that her soliciting behaviour does not appear to be one of the major components of the ewe's attractiveness. At present the factors leading to some ewes being more attractive to rams remain largely unknown, although the problem of male preference in sheep is one of economic relevance. For example, the chances of fertilization increase in the ewe with the number of services received (Mattner and Braden, 1967); ewes which attract a male only once or twice during their period of receptivity are likely to have a poorer chance of achieving conception.

The sexual preferences of the ewes were also studied by Key and MacIver (1980) by allowing them access to two tethered rams (one a Welsh Mountain ram and the other a Clun Forest ram). Six tests involved two mother-reared rams and two fostered rams. Overall the ewes showed significantly more interest in the mother-reared Welsh Mountain ram than in the Clun Forest mother-reared ram but more interest in the fostered Clun Forest ram than in the fostered Welsh Mountain ram. These numbers are, of course, too small to draw any generalizations except to point out that preferences were shown by the ewes.

It is generally assumed that dominant males will sire more offspring but as Dewsbury (1982) has pointed out, there is little firm evidence for this assertion, for even if the dominant males carry out most of the matings. Other factors, such as the differential fertilizing capacity of their semen, may affect the number of offspring sired by particular males. As Dewsbury maintains, one of the best examples that fully supports the hypothesis of dominance leading to more offspring sired, comes from a study by Guhl and Warren (1946). Using domestic cocks with different genetic markers they tested for any differential fertilizing capacities of the cockerels by artificially inseminating the hens with semen mixed from all the males. In one small flock the dominant males sired 52% of the chicks by normal matings but only 16% when his semen was mixed with that of the other males. In a second flock the results were less clear cut. However, the generality of this finding is doubtful, for in chickens it seems that even considering copulation rates alone, dominant males do not always prevail. For example, Kratzer and Craig (1980) investigated the effect of social rank on mating behaviour in cocks in flocks of different sizes and kept at different densities. Over the first five weeks of the experiment, the dominant males performed most of the completed matings and were less interrupted while mating than the other males. Nevertheless, the other cockerels, comprising 67% of the males in the experiment, later performed 58% of the completed matings.

An interesting effect of dominance on the mating behaviour of rams was reported by Lindsay *et al.* (1976). They assessed the relative dominance of 16 Merino rams using two criteria, competition over food and an oestrous ewe. Subsequently the effect of proximity of two relatively dominant or two relatively subordinate rams on the mating behaviour of each ram was examined. The ram that was being tested was placed in a central pen with an oestrous ewe and on either side of the pen were two pens in which the audience rams were placed. The sheep could see, smell and hear one another from one pen to another but could not make physical contact. In the control tests the rams were tested alone and the same ewes which were spayed and artificially brought into oestrus were used for all the males. Under these conditions, the dominant rams' rates of mounting and ejaculation were unaffected by the proximity of the subordinate rams, but both the mounting rates and ejaculate rates of the subordinate rams were significantly depressed by the proximity of the dominant rams, compared with their performances in the control tests.

6.1.3 Parental behaviour

In the mammalian farm species, maternal behaviour is of importance both under intensive and extensive farming conditions, for peri-natal mortality can lead to considerable economic losses. However, it is not only the behaviour of the dam that has to be considered but the whole process of mother–offspring bonding. It is essential that the newborn mammal obtain colostrum shortly after birth. In the case of the calf it is considered essential that an adequate amount be taken within the first six hours of life for it to obtain the maximum effect of the immunoglobulins. However, as we shall see, a number of factors work against this under modern agricultural conditions. Selman *et al.* (1970a,b) described the teat-seeking behaviour of neonatal calves and the dam's behaviour over the immediate post-natal period, using dairy heifers and cows as well as beef cows. In their sample of eight dairy cows, eight dairy heifers and ten beef cows, the last mentioned appeared to show the behaviour most conducive to the teat-seeking behaviour of the calves and to groom their calves more efficiently. The maternal behaviour of the dairy cows was generally described as less efficient and, furthermore, two of the dairy heifers were overtly aggressive towards their calves. The observations showed that the calf's teat-seeking behaviour was largely directed at the dam's fore or hind quarters and that there was an obvious preference on the calf's part to thrust its nose high and, if possible, into a recess such as the axilla or groin. The area where the calf concentrated this behaviour depended on the confirmation of the dam. When the udder and teats were situated at the highest part of the underbelly the behaviour was directed mainly towards the area of the udder but in others where the highest part of the underbelly was the xiphoid–axillary region, the behaviour was carried out often around the fore legs. If, in the latter type of female, pushing was round the udder the calf's nose was thrust high above the teats into the cow's groin. In multiparous dairy cows the udder is no longer ideal for this teat-seeking behaviour. Classifying cows into well or poorly shaped types before parturition, Selman *et al.* reported that calves found the teat significantly faster in the well shaped cows than in the cows with poor confirmation. In addition they reported that calves were sometimes seen to direct their teat-seeking behaviour towards a wall. It is apparent therefore that selection for high yielding milk cows has disturbed the normal innate behaviour of the calf, with the result that some calves do not get colostrum at the time most beneficial to them. The magnitude of this problem was shown in a study by Edwards (1982) who observed 161 dairy calves and found that 32% failed to suckle within the first six hours of life. As in the work by Selman and his colleagues, Edwards found the most important factor in determining the time to first suckling was the udder confirmation of the dam. In the case of cows having udders with the teats 50mm or more below the hock, 16 of the 34 calves did not suckle within six hours while in the calves of dams having udders with the teats 50mm above the hock, only three out of 20 did not suckle within 6 hours. Other factors affecting efficient teat-seeking behaviour were found to be abnormal maternal behaviour and the age of the dam, with the calves of older cows having difficulties, as well as the calves of heifers exhausted after parturition or those which sometimes rejected the calf. A sire effect was also found in that the calves of the different sires took different times to stand. A further factor was the temperature for as winter advanced and temperatures dropped the calves took longer to stand.

Mismothering can affect the colostrum intake of calves when cows are allowed to calve in groups. Cows approaching parturition are attracted to newborn calves and will often lick alien newly born calves. In another study Edwards (1983) found that 54% of the dams in a group of 39 licked alien calves and four of these abandoned their own calf when it was born in favour of the alien calf they had fostered before parturition. Of the calves born in this group, 82% directed their teat-seeking behaviour to both alien cows as well as their dams. In sheep, lambing in groups also causes problems with mismothering due to the same tendency for ewes approaching parturition to be attracted to newly born lambs.

As in the case of the cow, the ewe usually licks her lamb and it is thought that this enables the ewe to recognize her lamb and to discriminate against strange ones. Smith *et al.* (1966), on the evidence from very small numbers in experiments in which the lamb or lambs were removed from the ewe and strange lambs substituted, postulated that 20–30 minutes contact is necessary for the ewe to form a bond with her own lamb. Furthermore, there is a sensitive period in which she can develop this. Their work shows that ewes separated from their lambs for eight hours from birth and removed from the natal pen so as to avoid reinforcement from the amniotic fluids, readily accept their lambs. In seeking the teat the lamb appears to be guided by smoothness, warmth, a yielding substrate and olfactory cues from inguinal wax at the base of the udder (Vince *et al.*, 1984). However, teat-seeking behaviour does not appear to be so beset by problems in the lamb as it is in the case of the calf. Sheep nevertheless have additional problems. It is now a common practice to induce multiple births in ewes belonging to breeds that normally produce a single lamb per ewe, and such multiple births are not always accompanied by appropriate maternal behaviour. In one study in Australia on 103 fine-woolled Merino ewes with twins in large paddocks, 51 ewes became permanently separated from one lamb due to the ewe moving and leaving the lamb behind, mostly on the day of birth. The ewe was apparently unconcerned if followed by only one lamb, but separation

(a)

(b)

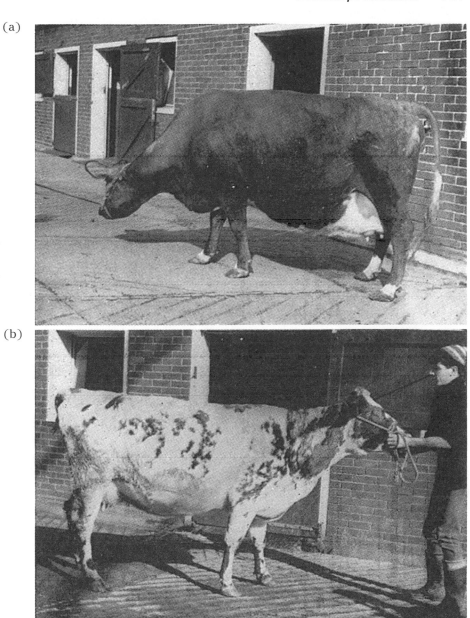

Figure 6.4 Examples of a cow with a poor shape (a) and another with a good shape (b). (From Selman *et al.*, 1970.)

from both lambs was a rare occurrence. Separation could not be accounted for by weakness of the lamb, for more than half of the lambs separated from the ewes were vigorous and mobile. In the study, 37.4% of the twins died as opposed to 9.6% of the singles (Stevens *et al.*, 1982). Another study in Australia and New Zealand (Alexander *et al.*, 1983) showed breed differences in this respect. At one research station in Australia, fine-woolled Merino were compared with Dorset Horn and Cross-breds (Border Leicester × Merino ewes × Suffolk rams) and data were also collected from a flock of Romney Marsh ewes in New Zealand. Although the ewes of the different breeds in the Autralian study were in different paddocks, the paddocks were essentially similar and the management and nutrition were the same. Under these conditions 46% of the Merinos with twins became permanently separated from a lamb while in the other two breeds, the Dorset Horn and the Cross-breds, the percentages were 17 and 0. Half of the cases of separation in the Merino could not be attributed to obvious precipitating factors such as weakness of lamb or difficult births. In the New Zealand pastures, which were topographically broken by gullies, and hillocks, 8% of the Romney Marsh ewes became permanently separated from a lamb, but some factor other than poor maternal behaviour could always be identified.

In the Australian data, it appeared that the longer the ewe remained at the birth site the less likely she was to lose a lamb through separation. During the course of these observations the lambs were tagged 1–12 hours after birth or, if born at night, as soon as possible the next day. About half the ewes of each breed moved 20–50m away, but returned as soon as the operator left. However, a significantly higher proportion of Dorset Horns remained with their lambs than Merino and Cross-breds. The behavioural trait was measured in New Zealand in 1050 ewes of six genotypes (breeds and breed crosses), three age groups and with litter sizes of 1–5 (O'Connor *et al.*, 1985). The ewes were scored on a five point scale with a score of one for fleeing at the approach of the shepherd and not returning, to five for staying close to the shepherd during the handling of her lamb or lambs. Within each litter size, survival from birth to weaning was found to increase the higher the ewe's protective behaviour score and this grew with the age of the ewe and with increasing litter size. No significant differences between genotypes were found. In summary, it was found that productivity in terms of kg/lamb weaned/ewe lambing increased from 11.2 for ewes with a score of 1, to 30.4 for ewes with a score of 5. From such results it is evident that this behaviour must be highly correlated with other traits that are beneficial to the ewes' offspring, but these have yet to be determined.

In the case of the domestic fowl the chicks are of course incubated and brooded artificially, but interest remains in the nesting behaviour of the hen. Certain breeding stocks are kept in pens and trap nesting is necessary to identify the eggs. Where hens may be kept on litter for commercial egg production, laying in the nests is necessary to prevent egg loss and the soiling of eggs. Laying on the floor is therefore a considerable problem economically and the whole behaviour pattern presents a welfare challenge (see Chapter 7). A considerable amount of work has been done on the underlying physiological mechanisms (Wood-Gush and Gilbert, 1964), and the nesting behaviour in connection with the laying of a particular egg is set in train on the previous day by hormones from the post-ovulatory follicle from which the egg is derived (Wood-Gush and Gilbert, 1973). With the ovulation of an egg, nesting behaviour the following day is thus inevitable. However, the fullness and adaptiveness of the behaviour is affected by the hens' environment and often raises questions concerning the welfare of the hen. In a battery cage the early stages of the behaviour in the absence of a good site and material gives way to frustration, while in a pen the hen may direct the later stages of nesting behaviour to nest building in the litter or direct her behaviour to the nest boxes provided (Wood-Gush, 1975). As stated, laying on the floor results in considerable economic losses and much research effort has gone into trying to find out the factors that would make an artificial nest attractive to the hens so that it might become a super-stimulus for all the hens. Further details of this approach are given in Chapter 7.

In many respects domestication seems to have had little effect on hen nesting behaviour. If allowed, the hen will build a nest not unlike the simple nests of many Gallinaceous birds and, in doing so, apparently pays attention to the site of the nest which in the wild would be highly adaptive. This emphasis on the nest site was shown in an experiment in which hens were kept in individual cages but released one at a time to lay in a pen with litter on the floor and nesting material such as feathers and string (Wood-Gush, 1975). Each hen was allowed to lay in the nest and then returned to her cage to be released to lay the next day in the pen. The second and later eggs of the clutch were laid in the same nest which was embellished by the hen. On occasions the nest was moved to another site in the pen after the hen had returned to her cage. When this was done, the hens on entering the next day to lay another egg in the clutch, built a new nest on the original site and rolled any eggs from the first nest in its new site to the nest in the former site. This conservatism also presents welfare problems, for even in a hen house with 4000 hens there was evidence of hens attempting to show it (Appleby *et al.*, 1986).

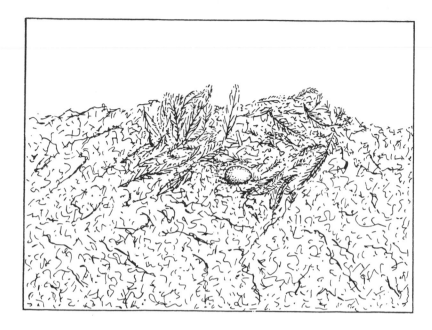

Figure 6.5 A nest built by a hen in a pen with wood shavings and feathers scattered on the floor. (From Wood-Gush, 1975.)

While the preferences of the hens may be largely guided by innate preferences for certain stimuli, learning can affect the birds' choice of a nest site. Appleby *et al.* (1983) reared some hens from hatching with the opportunity to perch and others without such opportunities. Four groups of 30 birds belonging to one strain were reared to five weeks in cages, then transferred to four pens. Two pens gave the birds the opportunity to perch and two did not. When the birds reached laying age each group was transferred to a pen with a block of six nest boxes in two tiers of three raised off the ground, so that the nests could be only approached by perches. During the first two weeks in the new pens, 15 and 17 hens from the pens without perching opportunities never perched. The percentage of eggs laid on the floor by the birds from the pens with no perching opportunities stabilized at 27 and 47%. The percentages from the hens with perching experience stabilized after a few days to almost zero and to 4% and considerable economic savings were thus achieved in terms of labour and soiled eggs.

6.1.4 Ingestive behaviour

Over the course of evolution, Natural Selection may be expected to have shaped a species' feeding behaviour so that the stimulus cues of important foods would be innately recognized in the form of visual or olfactory releasers by the ancestors of our modern farm livestock. In addition, early learning, such as imprinting, might be expected to occur in some cases. In agriculture, livestock have, probably since early times, been presented with novel foods, and domestication has certainly involved a greater readiness to accept novel food items, as has been shown in the domestic duck (Desforges and Wood-Gush, 1975). It might be expected therefore that learning about appropriate foods might be an important factor in the lives of farm animals. The pelleted diets commonly used bear little visual or olfactory resemblance to ancestral food items (see also Chapter 5) and even ingestion itself does not necessitate the use of some of the animal's innate feeding behaviour patterns; the battery cage hen, for example, does not have to scratch for her food and the pig does not have to root.

Scant attention has been paid to learning in this respect in farm animals. It is generally agreed in Animal Production that if the diet is nutritionally sound the animals will eat to develop and reproduce maximally. No doubts are even raised that if perhaps more was known about the animals' preferences and their development, even better records might be obtained. In the light of these attitudes relatively little work has therefore been directed towards the development of feeding behaviour in the common farm livestock species.

Two interesting studies however are reported here. One arose from a practical difficulty encountered in the sheep industry in Australia. Supplements in the form of whole grains, liquid molasses or compressed blocks of urea and molasses are often given when pasture conditions are poor. However, the intake of these supplements by mature animals is extremely variable and unpredictable. Lynch and his co-workers (Lynch *et al.*, 1983) investigated the intake of wheat by Merino lambs in relation to early experience. Lambs exposed to wheat at various ages from 1 week until 9 weeks of age showed little interest in wheat when hungry and offered it a week after weaning, whereas all those that had been exposed to it in the presence of their dams, which were used to it, ate appreciable amounts in identical tests. The second study by Key and MacIver (1980) has already been mentioned in connection with assortative mating. In addition, the grazing preferences of the sheep were studied. Lambs that had been born to, or fostered onto, Clun Forest ewes were kept with the ewes on a lowland pasture while those born to, or fostered onto, Welsh Mountain ewes were kept with the ewes on an upland pasture. At weaning the lambs were moved into two paddocks, each with an area of improved pasture and another area of unimproved hilly grasses and, later, observed for their grazing preferences. The observations, which were carried out in late summer and early autumn, showed that the Clun Forest-reared lambs preferred the improved pasture which was similar to their early paddock. The Welsh Mountain-reared lambs, on the other hand, showed no preference. Like the Australian study this shows that the feeding habits of the young sheep can be largely influenced by the habits of the dam, biological or fostered.

Under the conditions in which most farm livestock are kept it would not be surprising to learn that social factors are extremely important in motivating feeding behaviour, probably more so than the physiological factors in many circumstances. Csermely and Wood-Gush (1981) investigated the effect of auditory stimuli on the feeding behaviour of newly-weaned piglets. Having first established that there were periods when feeding was very infrequent, sounds of piglets suckling were played or a control sound during those periods when the piglets were between three and four weeks of age, over a total of 5 to 6 days. Playing of the suckling sounds led to a significant increase in ingestive behaviour compared with the control sound, and caused the piglets to feed at a time when they normally would not have fed. Usually, when piglets are weaned at 3 weeks of age and placed in a new environment, as these were, there is an arrest in their growth rates and the type of stimulation noted in this work might be beneficial in overcoming this arrest. However, it is not known from this experiment whether this stimulatory effect would merely alter the feeding pattern over a short time or whether it would lead to a real increase in food intake over the crucial period. Social facilitation of feeding behaviour may often be inferred from casual observations on farm animals, but such observations cannot generally rule out other causal mechanisms. Hsia and Wood-Gush (1984) demonstrated social facilitation in the pig in the following way. One fattening pig from a group of three placed in an individual stall was allowed to feed to satiation and then, when it had stopped feeding for nearly 10 minutes, one of its hungry group-members was introduced into the adjacent stalls where food was available. All the 'satiated' pigs ate on the introduction of a hungry group member. However it could not be certain that the first pig had really satiated itself initially, as isolation can depress the food intake of pigs (see below). A second experiment was therefore carried out in which two pigs of a group in adjacent stalls were allowed to feed. All the satiated pigs now resumed feeding on the introduction of the third hungry pig, thus clearly demonstrating the effect of social facilitation. Hsia and Wood-Gush (1983) also compared the effects of isolation, stall feeding and feeding in groups on the feeding behaviour of fattening pigs from small groups of four. Two levels of trough space were used in the group feeding, one allowing a generous amount of space per pig and the other allowing only one pig to feed at a time. The four treatments thus compared isolation, social facilitation and two levels of competition. Each pig went through each treatment according to a rigorous statistical design. At the end of each day the pigs slept in separate rooms so that energy expenditure during the night would be equalized between treatments. In three replicates, food consumption was found to be significantly greater in the conditions with light competition than in any of the others. The effect of stall feeding did not differ significantly from the condition of more severe competition and both these had higher scores than the score of pigs feeding alone.

Increasing on pig farms is the use of individual transponders worn on a collar around the sow's neck which allows her to enter a crate to feed. Several, and sometimes many, sows have to use the same crate (see Figure 6.6) which often leads to behavioural problems.

Figure 6.6 Sows with transponder collars at an automated feeder which allows one sow at a time to feed and which controls the amount of food for each individual sow. (By kind permission of Dr H. Simmons.)

The competitiveness of animals can be affected by early rearing conditions. Broom and Leaver (1978) reared two groups of six calves each using two rearing systems. Three calves of each group were reared in individual stalls and the other three as a group. It was later found that the calves reared in isolation were less competitive than the group-reared calves. Although the numbers are very small this experiment suggests that mixing calves from different types of rearing systems might place some at a relative disadvantage, particularly if resources such as concentrates are given in a limited space.

6.1.5 Housing and handling

As mentioned earlier, the design of housing systems has been influenced by a variety of factors and, generally, it has been expected that the animal will adapt itself to the housing supplied. As shown in Chapter 7, such an approach has led to a number of problems with regard to the welfare of the animals. Ideally the housing should be designed to fit not only the animals' bodily needs but also to fit the animals' behaviour so that social tensions and distress can be minimized. At the Edinburgh School of Agriculture an attempt has been made to do this in the case of pigs. Initially, observations were carried out on a small population of pigs kept under semi-natural conditions, with a minimum of restraint on their behaviour so as to be able to catalogue and analyse their behaviour (see Figure 7.4, Chapter 7).

Briefly, it was found (Stolba and Wood-Gush, 1984; Newberry and Wood-Gush, 1985, 1986, 1988) that the pigs formed small stable social groups which built communal nests for sleeping in at night, and that these nests were always far from the site where the pigs were fed. In addition these nests were built so as always to allow the pigs an open view on one side. In the morning, on rising, the pigs walked 4–11m to defaecate or urinate. Later defaecation during the day occurred on pathways going through bushes. Rooting was a common activity during the day and young pigs tended to lever small fallen tree trunks. Towards the end of the day the pigs carried nesting material to the communal nest or rearranged it. Most aggression occurred in connection with the concentrate food supplied, but if it was delivered in piles, suitably separated, then aggression declined significantly. When about to farrow the sows built individual nests well away from the communal nest and at sites which had bushes or a vertical surface at the back of the nest.

After farrowing the sows were aggressive for about a week but then often joined another closely related sow which had farrowed simultaneously and together they shared a nest. Synchronous oestrus was common amongst the sows. All showed lactational oestrus and conceived while lactating. Weaning of the piglets occurred at between 10 and 14 weeks. From these observations Stolba (1981) designed a housing system that incorporated features to allow the pigs to perform the behaviour outlined above. The plan view of the system is shown in Figure 6.7a. The basic social unit consists of four sows in four inter-leading pens. Piglets born into the system are kept there until sold so that there is no trauma of early weaning or the movement to a new environment. In each pen there is an area for the nest so that nests can be used communally or, at farrowing time, the sows can be shut off from one another. These nests which are separated from the feeding areas allow the pigs a good sight line. On rising in the morning pigs can walk about 9m to defaecate in a corridor that resembles a pathway. There is a rooting area with logs for levering and a rack in the pens contains hay for nesting building behaviour. More recently Kerr *et al.* (1988) have carried out some amendments, as shown in Figure 6.7b, to this system to cut down on the labour input. Over a three year period in these amended units, the reproductive perform-ance of the sows has been very good, the growth rate and food conversion rate of the piglets satisfactory. The only drawback from a commercial point of view has been high mortality amongst the newborn pigs, above average labour costs and, for the large-scale producer, an increased use of space. In all, the system has compared well with the average pig producer in terms of production and costs but is behind the most efficient producers with highly intensive units in terms of costs and profitability. Nevertheless, it is probable that features of the system will eventually find their way into conventional pig production systems. In the event of stringent welfare legislation, more features of the system will probably be widely adopted.

The handling of animals can be facilitated by a consideration of their behaviour and perceptual capabilities. Grandin (1980) has reviewed the current state of knowledge in this area of applied ethology. Illumination, natural or artificial, can be an important factor in affecting the movement of cattle and sheep which are reluctant to enter a dark area but will move towards a light area. Alternate dark and light stripes on the ground caused by shadow and light from fence paling will impede movement in pigs as well as cattle and sheep. The propensity of cattle, sheep and to a lesser extent of pigs to follow another animal can be used to ease movement from holding pens. For example, cattle and sheep will enter a shed or race through a narrow chute if they can see the animal ahead. In handling cattle or sheep it is important to consider the concept of flight distance in which the animal tries to keep a certain minimum distance from a human or

(a)

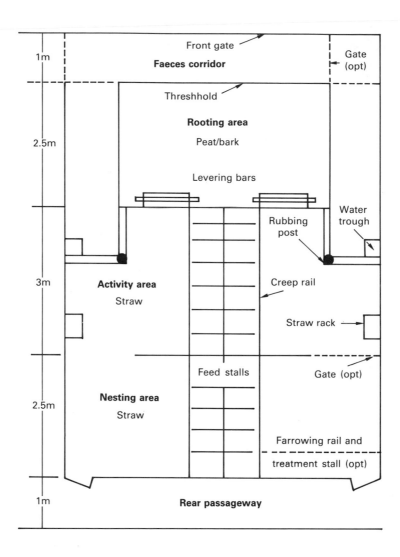

Figure 6.7 (a) Plan of original family pen. (From Kerr *et al.*, 1988.)
(b) Plan of modified family pen. (From Kerr *et al*, 1988.)
(c) Piglets and sow in the rooting area of an original family pen.

(b)

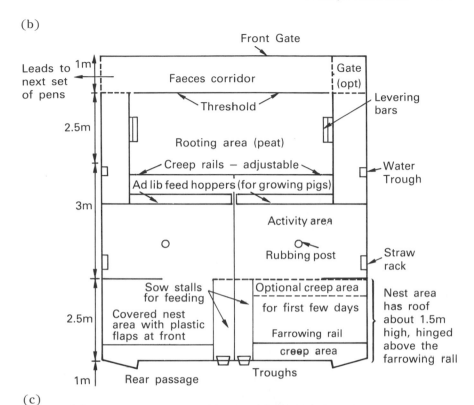

(c)

other animal. If infringement on the animal's space is too much the animal is then likely to panic, dash past or even become dangerous. The flight distance will vary between breeds and with the experience of the animals. Beilharz and Mylrea (1963) for example showed that in cattle the subordinate members of the herd tend to be nearer the handler than the dominant ones. In addition, cattle that have been housed will have shorter flight distances from humans than animals that have been kept on range. Positive reinforcement can also expedite handling under farm conditions (Huston, 1985).

Using the concept of flight distance, and bearing in mind the field of vision of cattle, Grandin describes how it is possible for a single person to herd a group of 5–20 cattle by the person positioning himself at a 45°–60° angle tangent to the shoulder of the leader and by moving will cause the leader to move and the rest to follow.

The postures of cattle can be fairly easily interpreted and are of potential use to the stockman. Bowes and Wood-Gush (1986) used them to interpret the movements of dairy cattle in their shed. Their study revealed that low ranking cows spent 20–30% of their active time in a submissive posture, while high ranking cows spent only 4% of their active time in that manner.

Conditioning can be used to ease the handling of stock and to reduce labour. Although conditioning is common on farms as, for example, in the use of various types of drinkers, it has not been very widely applied to ease handling and movement of stock. Kiley-Worthington and Savage (1978) successfully conditioned dairy cattle to walk 500–1500m to the milking parlour at the sound of an alarm bell, and thus saved the stockman an appreciable amount of time and effort. Using food rewards Hutson (1985) reported that sheep would readily go through a race to be clamped. However, if the sheep were clamped and inverted, the food rewards were less effective. In other words, the more severe the handling procedure, the less effective the reward.

Benign handling of dairy cows can lead to higher milk yields. Seabrook (1972) reported that in herds in which the cows approached the stockman, the yield per cow was significantly higher than in herds in which the cows lacked such confidence. The influence of three handling treatments on the behaviour, reproduction and free corticosteroid concentrations was studied in 15 male and 30 female pigs by Hemsworth *et al.* (1986). Two handling treatments, considered as pleasant and unpleasant, were imposed for 5 minutes three times a week from 11 weeks of age. The third handling treatment involved minimal contact with humans from that age. In a 3 minute test at 18 weeks, pigs in the pleasant treatment (gentle stroking on approaching

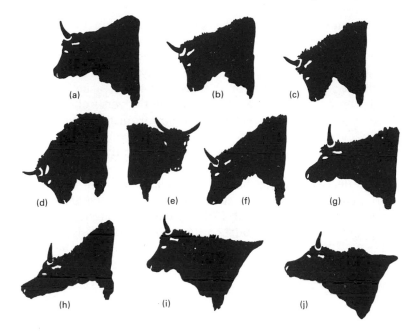

Figure 6.8 Cattle displays
(a) Normal position No. 1. The head in a neutral position.
(b) Lateral position No. 1. A milder form of threat. The animal places itself at a right-angle to its adversary.
(c) Lateral position No. 2. The form is very pronounced and the animal is close to combat. The animal places itself at a right-angle to its adversary.
(d) Combat position. This is a threat or stance for defence.
(e) Twisting of the head. A threat before or during a fight; the animal presents its side to its adversary.
(f) Normal position No. 2, browing. An expression of social inactivity.
(g) Approach position No. 1. A stance of greater assurance showing the intention to enter a group or make social contact (e.g. to play, fight, etc.).
(h) Approach position No. 2. A stance of less assurance often used by animals of lower social rank.
(i) Alert fighting position. The line of the back is concave and the base of the tail elevated. This is used by a defeated animal.
(j) Position of surveillance. This is a signal for the members of the group to be prepared for flight. The line of the back is concave and the back of the tail raised. (From Schloeth, 1958.)

the experimenter) approached the experimenter more quickly than those that had undergone the unpleasant treatment of being briefly shocked with a prodder (11v, 2mA). Gilts in the unpleasant treatment had a lower pregnancy rate at the second oestrus when mated to non-experimental boars than gilts in the group with pleasant treatment. Boars in the unpleasant treatment also showed adverse effects in their reproductive behaviour compared with boars from the pleasant treatment group. Furthermore, the free corticosteroid concentrations of the pigs in the unpleasant treatment were higher than those of the pigs in the pleasant treatment group. For many of the reproductive parameters the effect of minimal handling was intermediate between the other two treatments. However, it is uncertain as to what extent the differences between the pleasant and unpleasant treatments were due to the negative effects of the latter rather than the positive effects of the former.

6.1.6 Disease and behaviour

Sick animals often isolate themselves and behave in a listless manner. If they have localized pain they may lick the area. These are well known symptoms that inform the stockman that the animal is unwell, although they tell him little else. Some diseases, on the other hand, lead to very characteristic postures or gaits. Some behaviour patterns such as the possession of individual space, and the avoidance of faeces of conspecifics, will tend to decrease the risk of certain diseases, while other behaviour patterns such as the tendency to group at night will tend to increase the risk. The study of behaviour in relation to disease is an essential but under-researched area.

Parasites and their vectors cover a wide range of phyla, and ethology, together with the study of the behavioural ecology of the parasites, their vectors and hosts, can play an important role in the control of certain diseases. An interesting example of a recent ethological study in this area was concerned with outbreaks of bovine tuberculosis in parts of England. As it was considered that outbreaks of bovine tuberculosis in cattle were caused by infected badgers, Benham (1985) set out to investigate the relationship between a population of badgers and a herd of cattle that occupied the same area. Firstly, he investigated how cattle responded to grass turves that had been impregnated with faeces or urine of badgers. Although strong avoidance of these turves was found, some bites were taken and one cow showed no avoidance behaviour. In general, the results of these artificial tests suggested that infection is likely to occur in some other way. Direct contact in the field was also found to be an unlikely method of spreading the disease, for field observations showed that in the small sample of badgers studied there was avoidance of the cattle. Contact with possibly infected material was found to come about in an unexpected way: the cattle were found to rub their heads in the earth at the badgers' sett entrances. This behaviour elicited general interest in the herd and led to animals licking the ground and soil on each other. This suggests that, rather than extermination of badgers in unaffected areas as the Ministry of Agriculture had proposed, the fencing off of sets would be better.

6.1.7 Predation

In some areas predation of stock by carnivores can be more than an occasional nuisance. In the south western United States coyotes sometimes prey on lambs and Gustavson *et al.* (1976) attempted to use a behavioural technique to overcome the problem. Under laboratory conditions coyotes were presented with a choice between rabbits or chickens as prey. After it was established that rabbits were the preferred prey in all cases, each animal was presented with rabbit flesh tainted with lithium chloride, an emetic in small doses. After this treatment there was a complete decline in attacks on rabbits in the trials with a choice between rabbits or chickens. A field trial was followed in which sheep meat on a farm was similarly treated and following this it appeared that the lamb deaths due to killing by coyotes on this farm dropped by 30–60% compared with the previous three years. However, the effectiveness of this method of control on a larger scale has yet to be proven. (See also Chapter 3.)

6.1.8 Domestication of wild species

Recently there has been a move to domesticate wild species which it is felt would suit the conditions of certain areas better than the conventional farm livestock species. The Red deer (*Cervus elaphus* L) is being farmed in Britain and New Zealand (Blaxter *et al.*, 1974). In Africa some attempts have been made to farm some of the antelope species and in Central America efforts are being made to domesticate the paca, *Cuniculus paca* (Smythe, 1987). In some cases the process of domestication involves changing several aspects of the animal's behaviour. However, it is essential to realize that taming and domestication are different processes. In the Red deer, for example, while it is easy to capture young calves (Kelly and Drew, 1976) and to tame them sufficiently to be handled, their calves will be less amenable as adults unless also tamed by hand-rearing. Fortunately, early separation of the calf from the dam does not appear to cause long-term distress in the hind (Corrigall and Hamilton, 1977). Problems can also arise in relation to the keeping of the stags which become very vicious during the rutting season and which are difficult to keep in a herd (Suttie, 1985). In the paca, in addition to the problem of docility, there are other problems; they are not polygamous and there is much severe intra-specific aggression which can occur between the male and female at mating if the environmental conditions are not satis-factory and also between individuals over sleeping sites. However, the results from experiments to make the animals more tolerant of conspecifics look promising. From these examples, it can be seen that many of the problems found in domestication are behavioural and involve social and reproductive behaviour. However it is possible that ingestive behaviour may be involved in some cases.

6.2   MISCELLANEOUS APPLICATIONS

The pollination of many economic plants is under the control of animals (Faegri and van der Pijl, 1971; McGregor, 1976; Pesson and Louveaux, 1984). Insects are particularly important and among them bees are well known in this respect, apiculture being a long-standing sector of agriculture. A sound knowledge of the behaviour of these species to enhance their benefits and to protect them from disease is necessary.

Ethological knowledge can be, and is, used in several ways to control pests. Here we will discuss a number of pests, but the reader is directed to Chapters 2, 3, and 4 for information on avian pests. The brown rat is one of the most common pests of farm buildings. Be-havioural studies have shown that they display a high degree of neophobia in their feeding behaviour (Barnett, 1975). Any new food item is initially shunned. Only after several encounters with the new source of food will they begin to eat from it and, when they do, they tend to take very small amounts. This means that any bait should be presented many times, and only when it is obvious that large amounts have been consumed should the bait be poisoned. If the poison is added too soon, when the rats are taking only a small amount, they might only eat enough to make them ill, and then form a conditioned avoidance response to the bait.

The use of pheromones is practised in the control of certain insect pests, and some of these pheromones serve as releasers, acting on the nervous system to evoke a specific type of behaviour, such as attrac-tion of the sexes or aggregation or alarm behaviour. (See also Chapter

4.) Another class of pheromones, called primer pheromones, act physiologically to alter the endocrine and reproductive systems of the receptor animal. Such pheromones act either through physiological inhibition or enhancement (Matthews and Matthews, 1978). Pheromones acting as sex attractants include one used to survey and monitor the red-banded leaf roller (*Cargyrotaenia velutinana*) in apple orchards (Roelofs, 1975) and to control the California red scale (*Aonidiella aurantii*) in citrus orchards (Riehl *et al.*, 1980). In North America a pheromone that leads to aggregation has been used against the Scolytid bark beetles which attack trees. The first pioneer beetles settle on an injured or subnormal tree in response to volatile aldehydes or esters resulting from the abnormal enzyme activity of the injured tree. These pioneers then discharge a pheromone which attracts both sexes. The use of pheromones to control insect attacks has several features to commend it. They are more specific than insecticides which kill both harmful and beneficial insects indiscriminantly. Furthermore, since they are natural substances that regulate behaviour essential for survival, the insects are less likely to become resistant to them than to conventional insecticides. It should be remembered that some of the methods used against insects can be used against some other arthropod pests such as ticks. The use of the natural predators or parasites of the insect pest can be very effective (Huffaker and Smith, 1980). However, to be safe and effective the process requires an extensive knowledge about the behaviour and ecology of the species involved. The development of resistance to insecticides may sometimes involve behavioural adaptations. Broom (1981) cites a number of cases. Some houseflies and some malaria-carrying mosquitoes are resistant to DDT because they will not settle long enough on surfaces treated with DDT to absorb a lethal dose. Also the susceptibility of an insect species, or strain of species, may depend on the general level of activity of that species or strain. Conversely, one strain of grain weevil, which was resistant to the insecticide pyrethrum, was found to be less active, suggesting that its immunity is due to its members encountering less of the insecticide per unit of time.

It would be extremely erroneous to think that all crop pests are insects, for a number of nematode species are parasitic not only to plants but to animals as well. Dusenbery (1980) in a review of the studies on their behaviour points to the relative scarcity of such work. In a very limited number of species, the releasers that activate certain taxes and kineses have been investigated. As part of the soil fauna nematodes sometimes constitute most of the soil biomass (Nicholas, 1975). Often feeding on bacteria or fungi they occupy a key position in ecosystems by feeding on the primary decomposers of organic matter. In terms of trophic levels they play a major role at the second trophic level (Nicholas, 1975). Many other species are involved in soil dynamics (Wallwork, 1976). Indeed over 20 orders of arthropods inhabit the soil (Eisenbeis and Wichard, 1987). While the ecology of many of the species has been studied, their behaviour has not been investigated to the same extent. The only exception to this generalization are the Isoptera in which the behaviour of many species has been studied (Wilson, 1975). The role of earthworms in the soil has recently been reviewed by Lee (1985) but, again, as in many anthropod species the thrust of most work has been concerned with their ecology rather than their ethology.

6.3   CONCLUSIONS

From this review it can be seen that ethological studies can contribute much to agriculture not only to increase its efficiency, but to make it more humane and holistic. At the present time it would seem that agriculture is at a crossroads, either to become more intensive and to produce more food from less land and with less labour, or to become more ecologically orientated. In either event it would seem that ethology will have a significant role.

REFERENCES

Alexander, G., Stevens, D., Kilgour, R. *et al.* (1983) Separation of ewes from twin lambs: Incidence in several breeds. *Appl. Anim. Ethol.,* **10**, 301–17.

Anderson, M., Schaar, J., Wiktorsson, H. (1984) Effects of drinking water flow rates and social rank on performance and drinking behaviour of tied-up dairy cows, *Livestock Prod. Sci.,* **11**, 599–610.

Appleby, M.C. (1983) The possibility of linear hierarchies. *Anim. Behav.,* **31**, 600–8.

Appleby, M.C., McRae, H.E., and Duncan, I.J.H. (1983) Nesting and floor-laying by domestic hens: Effects of individual variation in perching behaviour. *Behav. Anal. Letters,* **3.**, 345–52.

Appleby, M.C., Maguire, S.N. and McRae, H.E. (1986) Nesting and floor laying by domestic hens in a commercial flock. *Brit. Poult. Sci.,* **27**, 75–82.

Arave, C.W. and Albright, J.L. (1976) Social rank and physiological traits of dairy cows as influenced by changing group membership.

J. Dairy Sci., **59**, 974–81.

Barnett, S.A. (1975) *The Rat: A study in behaviour*, Univ. Chicago Press, London.

Barton, E.P., Donaldon, S.L., Ross, M. and Albright, J.L. (1974) Social rank and social index as related to age, body weight and milk production in dairy cows, *Proc. Indiana Acad. Sci.*, **83**, 473–7.

Beilharz, R.G., Butcher, D.R. and Freeman, A.E. (1966) Social dominance and milk production in Holsteins. *J. Dairy Sci.*, **49**, 887–92.

Beilharz, R.C. and Mylrea, P.J. (1963) Social position and the behaviour of dairy heifers in yards. *Anim. Behav.*, **11**, 522–33.

Benham, P.F.J. (1985) The behaviour of badgers and cattle and some factors that affect the chance of contact between the species. *Appl. Anim. Behav. Sci.*, **14**, 390–1.

Blaxter, K.L., Kay, R.N.B., Sharman, G.A.M. *et al.* (1974) *Farming the Red Deer*, H.M.S.O. Edinburgh.

Bowes, K. and Wood-Gush, D.G.M. (1986) Social tension in dairy cows. *Appl. Anim. Behav. Sci.*, **16**, 95–6.

Broom, D.M. (1981) *Biology of Behaviour*, Cambridge Univ. Press, Cambridge.

Broom, D. and Leaver, J.D. (1978) Effects of group rearing or partial isolation on later social behaviour of calves. *Anim. Behav.*, **26**, 1255–63.

Collis, K.A. (1976) An investigation of factors related to the dominance order of a herd of dairy cows of similar age and breed. *Appl. Anim. Ethol.*, **2**, 167–73.

Corrigall, W. and Hamilton, W.J. (1977) Reaction of red deer (*Cervus elaphus L*) hinds to removal of their suckled calves for hand rearing. *Appl. Anim. Ethol.*, **3**, 47–55.

Csermely, D. and Wood-Gush, D.G.M. (1981) Artificial stimulation of ingestive behaviour in early weaned pigs. *Biol. Behav.*, **6**, 159–66.

Desforges, M.F. and Wood-Gush, D.G.M. (1975) A behavioural comparison of domestic and mallard ducks: habitation and flight reactions. *Anim. Behav.*, **23**, 692–7.

Dewsbury, D.A. (1982) Dominance rank, copulatory behaviour and differential reproduction. *Quart. Rev. Biol.*, **57**, 135–57.

Dickson, D.P., Barr, G.R. and Wieckert, D.A. (1967) Social relationships of dairy cows in a feed-lot. *Behaviour*, **29**, 195–203.

Dusenbery, D.B. (1980) Behaviour of free-living nematodes, in *Nematodes as Biological Models*, Vol. I (ed. B.M. Zuckerman), pp. 127–58, Academic Press, London.

Edwards, S.A. (1982) Factors affecting the time to first suckling in dairy calves. *Anim. Prod.*, **34**, 339–46.

Edwards, S.A. (1983) The behaviour of dairy cows and their newborn calves in individual or group housing. *Appl. Anim. Ethol.*, **10**, 191–8.

Eisenbeis, G. and Wichard, W. (1987) *Atlas on the Biology of Soil Arthropods* (Translated by E.A. Mole) pp. 437, Springer Verlag, Ijmuiden.

Ewbank, R. and Meese, G.B. (1971) Aggressive behaviour in groups of domesticated pigs on removal and return of individuals. *Anim. Prod.*, **13**, 685–93.

Faegri, K. and Van Der Pijl, L. (1971) *The Principles of Pollination Ecology*, 2nd edn, Pergamon Press, Oxford, pp. 291.

Friend, T. and Polan, C.E. (1978) Competition order as a measure of social dominance in diary cattle. *Appl. Anim. Ethol.*, **4**, 61–70.

Geist, V. (1971) *Mountain Sheep: A Study in Behaviour and Evolution*, University of Chicago Press, London.

Grandin, T. (1980) Livestock behaviour as related to handling facility design. *Internat. J. for Study of Anim. Prob.*, **1**, 33–52.

Grubb, P. and Jewell, P.A. (1966) Social grouping and home range in feral Soay sheep. *Symp. of Zool. Soc. Lond.*, **18**, 179–210.

Guhl, A.M. and Warren, D.C. (1946) Number of offspring sired by cockerels related to social dominance in chickens. *Poult. Sci.*, **25**, 460–72.

Gustavson, C.R., Kelly, D.J., Sweeney, M.M. and Garcia, J. (1976) Prey-lithium aversions 1: Coyotes and Wolves. *Behav. Biol.*, **17**, 61–72.

Hall, S.J.G. (1979) Studying the Chillingham wild cattle. *Ark*, **6**, 72–6.

Hemsworth, P.H., Barnett, J.L. and Hansen, C. (1986) The influence of handling by humans on the behaviour, reproduction and corticosteroids of male and female pigs. *Appl. Anim. Behav. Sci.*, **15**, 303–14.

Hemsworth, P.H., Beilharz, R.F. and Galloway, D.B. (1977) Influence of social conditions during rearing on the sexual behaviour of the domestic boar. *Anim. Prod.*, **24**, 245–51.

Hemsworth, P.H., Findlay, J.K. and Beilharz, R.G. (1978) The importance of physical contact with other pigs during rearing on the sexual behaviour of the male domestic pig. *Anim. Prod.*, **27**, 201–7.

Hsia, Liang Chou and Wood-Gush, D.G.M. (1983) A note on social facilitation and competition in the feeding behaviour of pigs. *Anim. Prod.*, **37**, 149–52.

Hsia, Liang Chou and Wood-Gush, D.G.M. (1984) Social facilitation in the feeding behaviour of pigs and the effect of rank. *Appl. anim. Ethol.*, **11**, 265–70.

Huffaker, C.B. and Smith, R.F. (1980) Rationale, organisation and development of a natural integrated pest management project in *New Technology of Pest Control* (ed. C.B. Huffaker) John Wiley and Sons, Chichester.

Hughes, B.O. (1977) The absence of a relationship between egg production and dominance in caged laying hens. *Brit. Poult. Sci.,* **18**, 611–16.

Hutson, G.D. (1985) The influence of barley food rewards on sheep movement through a handling system. *Appl. Anim. Behav. Sci.,* **14**, 263–73.

Kelly, R.W. and Drew, K.R. (1976) Shelter seeking and sucking behaviour of the red deer calf *Cervus elaphus* in a farmed situation. *Appl. Anim. Ethol.,* **2**, 101–11.

Kerr, S.C.G., Wood-Gush, D.G.M., Moser, H. and Whittemore, C.T. (1988) Enrichment of the production environment and enhancement of welfare through the use of the Edinburgh family pen system of pig production. *Research and Development in Agriculture,* **5**, 171–86.

Key, C. and MacIver, R.M. (1980) The effects of maternal influences on sheep: breed differences in grazing, resting or courtship behaviour. *App. Anim. Behav.,* **6**, 33–48.

Kiley-Worthington, M. and Savage, P. (1978) Learning in dairy cattle using a device for economical management of behaviour. *Appl. Anim. Ethol.,* **4**, 119–24.

Konggaard, S.P. and Kohn, C.C. (1976) Investigations concerning feed intake and social behaviour among group fed cows. Under loose housing conditions, II. Factors influencing the individual intake of grass silage fed *ad-libitum. Beretning fra Statens Husdyrbrugsforsøg,* **441**, 1–26.

Kratzer, D.D. and Craig, J.V. (1980) Mating behaviour of cockerels: effects of social status, group size and group density. *Appl. Anim. Ethol.,* **6**, 49–62.

Krohn, C.C., and Konggaard, S.P. (1979) Effects of isolating first-lactation cows from older cows. *Livestock Prod. Sci.,* **6**, 137–46.

Kronsniunas, E.H. (1984) Social facilitation and foraging behaviour of the feral pig (*Sus Scrofa*) on Santa Cruz Island, California. M.A. Thesis, Univ. of California, Davis.

Lee, K.E. (1985) *Earthworms. Their ecology and relationships with soils and land use.* Academic Press, London. pp. 411.

Lill, A. and Wood-Gush, D.G.M. (1965) Potential ethological isolating mechanisms and assortative mating in the domestic fowl. *Behaviour,* **25**, 16–44.

Lindsay, D.R., Dunsmore, D.E., Williams, J.D. and Syme, G.J. (1976) Audience effects on the mating behaviour of rams. *Anim. Behav.,* **24**, 818–21.

Lynch, J.J., Keogh, R.G., Elwin, R.L. *et al.* (1983) Effects of early experience on the post-weaning acceptance of whole grain wheat by

Fine-wooled Merino lambs. *Anim. Prod.*, **36**, 175–83.

McBride, G. (1962) The interactions between genotypes and housing environments in the domestic hen. *Proc. Aust. Soc. Anim. Prod.*, **4**, 95–102.

McBride, G. (1964) Social behaviour of domestic animals. Effect of the peck order on poultry productivity. *Anim. Prod.*, **6**, 1–7.

McBride, G., James, J.W. and Hodgens, N. (1964) Social behaviour of domestic animals. IV. Growing pigs. *Anim. Prod.*, **6**, 129–39.

McBride, G., James, J.W. and Wyeth, G.S.F. (1965) Social behaviour of domestic animals VII. Variation in weaning weight in pigs. *Anim. Prod.*, **7**, 67–74.

McBride, G., Parker, L.P. and Foenander, F. (1969) The social organization and behaviour of the feral domestic fowl. *Anim. Behav. Monogr.*, **2**, 172–81.

McGregor, S.E. (1976) Insect pollination of cultivated crop plants. Agric. Handbook No. 496 Agric. Res. Service U.S.D.A. Washington, DC.

Matthews, R.W. and Matthews, J.F.R. (1978) *Insect Behaviour*, John Wiley and Sons, Chichester.

Mattner, P.E. and Braden, A.W.M. (1967) Studies in flock mating of sheep 2. Fertilization and prenatal mortality. *Aust. J. Exp. Agric. Anim. Husb.* **7**, 110–16.

Newberry, R.C. and Wood-Gush, D.G.M. (1985) The suckling behaviour of domestic pigs in a semi natural environment. *Behaviour*, **95**, 11–25.

Newberry, R.C. and Wood-Gush, D.G.M. (1986) Social relationships of piglets in a semi-natural environment. *Anim. Behav.*, **34**, 1311–14.

Newberry, R.C. and Wood-Gush, D.G.M. (1988) The development of certain behaviour patterns in piglets under semi-natural conditions. *Anim. Prod.*, **46**, 103–9.

Nicholas, W.L. (1975) *The Biology of Free-living Nematodes*, Clarendon Press, Oxford.

O'Connor, C.E., Jay, N.P., Nicol, A.M. and Beatson, P.R. (1985) Ewe maternal behaviour score and lamb survival. *Proc. N.Z. Soc. Anim. Prod.*, **45**, 159–62.

Pesson, P. and Louveaux, J. (1984) Pollinisation et productions végétales, INRA. Paris.

Price, E.O. (1984) Behavioural aspects of animal domestication. *Quart. Rev. Biol.*, **59**, 1–32.

Riehl, L.A., Brooks, R.F., McCoy, C.W. *et al.* (1980) Accomplishments toward improving integrated pest management for citrus, in *New Technology of Pest Control* (ed. C. Huffaker) John Wiley and Sons, Chichester, pp. 319–64.

Roelofs, W.L. (1975) Insect communication — chemical, in *Insects, Science and Society* (ed. D. Pimental) Academic Press, New York, pp. 79–99.

Sambraus, H.H., Fries, B. and Osterkorn, K. (1978/79). Das Sozialgeschehen in einer Herde hornloser Hochleistungsrunder. *Z. Tierzücht. Züchtungsbiol*, **95**. 81–8.

Sambraus, H.H., Osterkorn, K. and Kräusslich (1979) Beziehungen zwischen sozialem Rang und Milchleistung in einer Hochleistungsherde. *Züchtungskunde*, **51(4)**, 289–92.

Schein, M.W. and Fohrman, M.H. (1955) Social dominance relationships in a herd of dairy cattle, *Brit. J. Anim. Behav.*, **3**, 45–55.

Schoeth, R. (1985) Cycle annuel et comportement social du taureau de Carmargue, *Mammalia*, **22**, 121–39.

Seabrook, M.F. (1972) A study to determine the influence of the herdman's personality on milk yield. *J. Agric. Labour Sci.*, **1**, 45–59.

Selman, I.E., McEwan, A.D. and Fisher, E.W. (1970a) Studies on natural suckling in cattle during the first eight hours post partum. I. Behavioural Studies (Dam). *Anim. Behav.*, **18**, 276–83.

Selman, I.E., McEwan, A.D. and Fisher, E.W. (1970b) Studies on natural suckling in cattle during the first eight hours post partum. II. Behavioural Studies (Calves). *Anim. Behav.*, **18**, 284–9.

Smith, F.V., van Toller, C. and Boyes, T. (1966) The 'Critical Period' in the attachment of lambs and ewes. *Anim. Behav.*, **14**, 120–5.

Smythe, N. (1987) The paca (*cuniculus paca*) as a domestic source of protein for the neo-tropical, humid lowlands. *Appl. Anim. Behav. Sci.*, **17**, 155–70.

Soffie, M., Thines, G. and Marneffe, G. (1976) Relation between milking order and dominance value in a group of dairy cows. *Appl. Anim. Ethol.*, **2**, 271–6.

Stevens, D., Alexander, G., and Lynch, J.J. (1982) Lamb mortality due to inadequate care of twins by Merino ewes. *Appl. Anim. Behav.*, **8**, 243–52.

Stolba, A. (1981) A family system of pig housing, in Symposium on Alternatives to Intensive Husbandry Systems, UFAW, Potters Bar, Herts. pp. 52–67.

Stolba, A. and Wood-Gush, D.G.M. (1984) The identification of behavioural key features and their incorporation into a housing design for pigs. *Annales de Recherches Veterinaires*, **15**, 287–9.

Suttie, J.M. (1985) Social dominance in farmed red deer stags. *Appl. Anim. Behav. Sci.*, **14**, 191–9.

Syme, G.J. (1974) Competitive orders as measures of social dominance. *Anim. Behav.*, **22**, 931–40.

Syme, G.J. and Syme, L.A. (1979) *Social Structure in Farm Animals.*

Developments in animal and veterinary services 4, Elsevier, Amsterdam.

Tilbrook, A.J. and Lindsay, D.R. (1987) Differences in the sexual 'attractiveness' of oestrus ewes to rams. *App. Anim. Behav. Sci.*, **17**, 129–38.

Vince, M.A., Ward, T.M. and Reader, M. (1984) Tactile stimulation and teat sucking behaviour in newly born lambs. *Anim. Behav.*, **32**, 1179–84.

Wallwork, J.A. (1976) *The Distribution and Density of Soil Fauna*, p. 355, Academic Press, London.

Wilson, E.O. (1975) *Sociobiology. The new synthesis*, The Belknap Press of Harvard Univ. Press, London.

Wood-Gush, D.G.M. (1954) The courtship of the Brown Leghorn cock. *Anim. Behav.*, **2**, 95–102.

Wood-Gush, D.G.M. (1955) The behaviour of the domestic chicken: A review of the literature. *Brit. J. Anim. Behav.*, **3**, 81–110.

Wood-Gush, D.G.M. (1975) Nest construction by the domestic hen: Some comparative and physiological considerations, in *Neural and Endocrine Aspects of Behaviour in Birds* (eds P. Wright, P.G. Caryl and D.M. Vowles), pp. 35–49, Elsevier, Amsterdam.

Wood-Gush, D.G.M. and Gilbert, A.B. (1964) The control of the nesting behaviour of the domestic hen II. The role of the ovary. *Anim. Behav.*, *12*, 451–3.

Wood-Gush, D.G.M. and Gilbert, A.B. (1973) Some hormones involved in the nesting behaviour of hens. *Anim. Behav.*, **21**, 98–103.

Part Three
Animal Welfare

7

Promoting the welfare of farm and captive animals

IAN J.H. DUNCAN and TREVOR B. POOLE

7.1 WHAT IS ANIMAL WELFARE?

One of the problems associated with animal welfare is in deciding exactly what it means. 'Welfare' is a common word of everyday speech that means different things to different groups of people. The Oxford English Dictionary defines 'welfare' as 'The state or condition of doing or being well; good fortune, happiness or well-being; thriving or successful progress in life, prosperity'. Are these definitions appropriate when considering animal welfare? The first and last definitions convey the meaning of being healthy or thriving, and so could be applied quite unequivocally to animals (or indeed to plants). Welfare would thus be synonymous with health. However, when applied to human beings, welfare means more than this, encompassing mental health as well as physical health. Moreover, the word 'happiness' in the second definition implies a pleasant subjective emotional experience. These facts have been taken into account in the descriptions of animal welfare which follow. The Brambell Committee, which was set up by the British Government to investigate the welfare of intensively-housed livestock, stated that 'Welfare is a wide term that embraces both the physical and the mental well-being of the animal. Any attempt to evaluate welfare, therefore, must take into account the scientific evidence available concerning the feelings of animals that can be derived from their structure and function and also from their behaviour' (Command Paper 2836, 1965). Hughes (1976a) thought that 'Welfare is a state of complete mental and physical health, where the animal is in harmony with its environment'. More recently Carpenter (1980) declared that 'The welfare of managed animals relates to the degree to which they can adapt without suffering to the environments designated by man. So long as a species remains within the limits of the environmental range to which it can adapt, its well-being is assured.' Duncan and Dawkins (1983) considered these descriptions and decided that it was impossible to give welfare an exact and un-ambiguous definition. A broad working definition would be one which contained the main ideas from the foregoing descriptions, namely, 'the animal in physical and mental health', 'the animal in harmony with its environment', 'the animal adapting without suffering' and it should also take account of 'the animal's feelings'.

Similarly, it is probably impossible to give a precise scientific defini-tion to 'suffering'. The Brambell Committee (Command Paper 2836, 1965) attempted to overcome this difficulty by drawing up a list of states such as pain, exhaustion, fright and frustration which, they claimed, would be accompanied by unmistakable signs of suffering. However, it is more difficult to decide whether or not animals experi-ence other states of suffering, such as boredom, loneliness and grief, which are experienced by human beings. There is also the possibility that animals may experience certain states of suffering that are not experienced by human beings. For example, it is difficult for us to imagine what a captive migratory bird must experience when the time comes from it to migrate and it cannot. This method of defining 'suffering' therefore has the disadvantage that some states of suffering may be missed. For this reason, it is probably safer to give 'suffering' a broad working definition, such as that adopted by Dawkins (1980), namely, 'a wide range of unpleasant emotional states'.

Although physical health and freedom from injury are important, ultimately, it is how the animal 'feels' about its bodily state, how it 'perceives' its environment and how 'aware' it is of these feelings and perceptions that are crucial for its welfare. 'Feeling' is sensing bodily events and 'perceiving' is detecting and interpreting signals that normally originate in external events. An animal 'is aware of' (or notices) a stimulus if it feels it (for internal events) or perceives it (for external events). These processes of feeling, perceiving and being aware are the simplest of the cognitive processes and their possession is probably limited to the vertebrates (although there is some debate as to whether the higher invertebrates, and in particular the cephalo-pods, have simple cognitive abilities). In general, not only do the species kept in captivity by man tend to be vertebrates, but they tend to be higher vertebrates, namely birds and mammals (though see Chapter 5), the very groups which are thought to have the greatest cognitive abilities (Walker, 1983) and therefore the groups whose welfare is most at risk.

Another problem associated with animal welfare is that it lies at the intersection of science, ethics and aesthetics. Thus, although it may be possible for a scientist to measure how painful a particular procedure may be to an animal or how frustrated it may be when confined in a certain environment, the decision as to whether or not that procedure or that environment should be allowed is an ethical decision and therefore one that should be made by society in general. These decisions usually involve some balancing of costs and benefits. Thus a laboratory experiment which involved frustrating some tens of hens might be justified if it resulted in the design of a commercial husbandry system which eliminated frustration for millions of hens. On the other hand, an experiment which involved subjecting tens of monkeys to severe pain might not be justified if the only benefit was a small advancement in learning theory. However, not all ethical decisions are so clear cut and Driscoll and Bateson (1988) have set out some guidelines to help in these decisions. Similar ethical decisions have to be made with regard to the other ways in which we exploit animals. Is a particular farming practice justified if it leads to the animals being hungry for 30% of the time but the final product being 10% cheaper? Is the existence of zoos justified if some of the animals in them suffer from severe frustration but they result in certain species being conserved and in an increased public awareness of endangered species? Are we justified in keeping pet dogs which may be denied social contact for long periods in the day but which bring us great enjoyment when we are with them? The scientist, and particularly the behavioural scientist, may be able to give an increasing amount of information on the costs and benefits involved in these decisions, but the decisions themselves are ethical ones which should be made by society in general.

The matter may be further complicated by aesthetic considerations. Thus it may offend some people's aesthetic sense to see animals behind bars no matter what the actual effects of bars on the welfare of the animals behind them.

7.2   HISTORICAL BACKGROUND

The recent public concern about the welfare of captive animals is probably the result of a growing realization that the higher animals may be capable of experiencing suffering, in a similar manner to human beings. The controversy has been brought to a head by the circumstances of animals kept under intensive conditions for farming purposes and the plight of some animals kept for scientific research. However, it should be remembered that all captive animals are, by definition, subjected to an environment that is 'un-natural', in that it is different in some way from the environment in which the progenitors of the captive animals evolved. The problem has not, therefore, been confined to recent times. It is likely that, in the past, many husbandry practices led to reduced welfare. For example, it was common practice in the 19th and early 20th century to keep herds of dairy cattle in byres in the middle of cities. These cows spent most of their lives tied by the neck in stalls in a very artificial and crowded environment. Heavy horses, which were worked extremely hard for six days a week and then tied up in an environment with no stimulation whatsoever on a Sunday, developed all sorts of maladaptive behaviour patterns. Many forms of livestock were given no protection from the weather during the winter and barely enough food to survive. Menageries contained various species of exotic wild animals in totally unsuitable bare cages.

In modern times, poor welfare has been brought about in a number of different ways. There has been a development of husbandry systems for agricultural species which appear to reduce welfare. For example, these systems often provide a very inappropriate physical environment which may frustrate certain behaviour patterns, lack essential stimuli, or provide too much or too little general stimulation. In addition, these systems often keep animals in an inappropriate social environment, with the group size being too large or too small, the population density being too great and the social mix being unsuitable for the animals concerned. These systems include keeping laying hens in small cages with a sloping floor and automating all husbandry procedures (Figure 7.1), keeping breeding sows tethered in individual stalls throughout pregnancy, keeping groups of fattening pigs at high density in very barren concrete pens at low levels of illumination and keeping veal calves in small individual crates with no bedding. It is ironic that the original reason for the development of many of these systems was to improve hygiene (the battery cage separates the bird from its faeces and so reduces the risk of re-infection with many organisms) or to reduce mortality (the tether-stall system ensures that each pregnant sow in a herd gets exactly the rations it requires without fighting with other sows), improvements that should, on the face of it, *increase* welfare. However, these developments paid little or no heed to the behaviour of the animals involved, and the overall effect has been an increase in stress due to frustration or fear or crowding, albeit with an increase in hygiene or decrease in mortality.

Probably the biggest cause of reduced welfare today is associated with the handling, transportation and pre-slaughter management of animals destined to be used as human food. The problem is a huge one. For example, within the European Community, about eleven million broiler chickens are transported to slaughter each working day (derived from Flock, 1986) and between 10% and 30% of these are

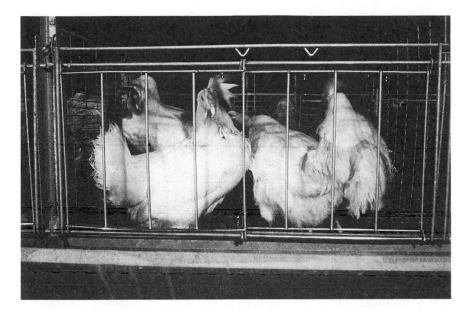

Figure 7.1 Laying hens in a battery cage.

injured during the process (Gerrits *et al.*, 1985). The welfare of these injured birds will be impaired (by definition) and it is likely that the remainder will suffer varying degrees of stress. Although much can be done to improve handling and transport facilities (Grandin, 1983), the question must be asked if it is necessary to transport all this food as live animals, and if slaughter on the farm would not be a much more humane solution.

Another recent source of poor welfare has been the conditions under which laboratory animals are kept. Apart from the stressful procedures to which they are sometimes exposed, these animals are often kept in environments that are less than ideal. Once again the reasons are easy to understand; the requirements of the research in question often means that animals are kept in extremely hygienic conditions (medical and veterinary research) or uninfluenced by other animals (psychological research). This has led to animals being kept in conditions that are psychologically as well as biologically sterile and often in social isolation. As with the agricultural species, the overall effect has been an increase in stress in these animals which, apart from the deleterious effect on their welfare, may cast doubts on the validity of the research results.

In many respects zoos have led the way in paying more attention to the welfare of animals in their care. Ever since Hediger (1950) described the abnormal behaviour that was often exhibited by wild animals in captivity, progressive zoos have tried to take account of the natural history of species when designing their enclosures. However, it would be a mistake to think that all the problems had been solved. Animals are still to be seen in bare cages or in enclosures which give the impression to the public of environmental richness without actually catering for the needs of the species in question. For example, if canids such as wolves and dingos are not provided with huge areas over which to patrol, they develop very stereotyped pacing behaviour over a few routes in their enclosure which is suggestive of reduced welfare. Also, the primates and particularly the great apes require not only a very rich environment but a constantly changing one which will occupy their almost insatiable tendency to explore.

In the examples given so far, poor welfare has been the result of some deficiency in the captive environment. However, there is another way in which man has interfered with some species to the detriment of their welfare and that is by changing them genetically. It appears that sometimes intensive selection pressure for a particular trait can result in an animal for which a satisfactory environment cannot be designed (or at least has not been designed to date). For example, the breeding stock for fattening pigs, turkeys and broiler chickens have to be kept constantly hungry in order to prevent them from becoming so obese that they cannot breed. Selection for weight gain in turkeys and broiler chickens has resulted in a huge increase in the incidence of various painful orthopaedic diseases (Duff and Hocking, 1986; Duff *et al.*, 1987). With regard to research animals, mice are now available that have been produced using genetic engineering techniques, which are guaranteed to develop cancer within 90 days (Murphy, 1988) and whose welfare is therefore also guaranteed to be reduced, no matter in what environment they are kept.

BEHAVIOURAL NEEDS

The importance of housing animals in 'appropriate' environments has been emphasized in the previous section, but what is an 'appropriate' environment? A simple-minded view might be that an environment would be ideal if it provided every 'goal' that the animal could possibly need. By 'goal', is meant being able to behave socially, sexually and so on, as well as such things as food, water and thermal comfort. There would probably be general agreement that there are certain behaviour patterns that captive animals do not 'need' to express, although these might be part of their total behavioural repertoire. The most straight-forward examples of behavioural patterns of this type would be anti-predator responses; generally speaking, there is no need for captive animals to perform anti-predator responses and one would not regard their absence as being indicative of reduced welfare. However, there is one very important proviso, which is that the stimuli eliciting these responses should *never* occur in the captive environment. If there is any chance that the responses could be triggered either by a predator or by some sign stimulus connected to a predator, then the captive environment should be designed to allow the responses to occur in a non-damaging way. For example, laying hens in battery cages are usually housed in controlled environment houses where there is no chance of them coming in contact with a ground predator. However, some farmers use soft bristle brushes to remove dust from the cages and these brushes cause extreme alarm in the hens because either the bristles or the brush movement simulates a ground predator. The birds cannot make the appropriate fleeing response in cages and so suffer extreme fear and sometimes physical damage when cage dusting takes place.

Most behaviour patterns are not elicited entirely by external stimuli but have factors which act from within the animal. For example, feeding behaviour is motivated largely by internal factors which build up with time since the animal last fed; female mammalian sexual behaviour is motivated by internal factors which fluctuate with time (the oestrus cycle). It is with this type of behaviour pattern that the question of 'needs' and 'goals' arises. A major problem in this area is in identifying exactly what it is that the animal finds rewarding. For example, with regard to food, is it the feeling of satiation that is rewarding or might not the actual act of eating have some reward in itself? Put simply, is it the goal or moving towards the goal that is important? Introspection tells us that the latter is a possibility and in fact there is experimental evidence that, in rats, food ingested by way of the mouth is more rewarding than a similar quantity given through a stomach fistula (Miller and Kesson, 1952). Notwithstanding this evidence, Baxter (1983) has argued that if all the animal's environ-mental requirements could be fully met by appropriate environmental design, then motivation to perform the behaviour in question would not be triggered. In the case of a pre-partum sow, he postulated that if all the functional requirements of a perfect nest were met in terms of temperature control, comfort, and so on, then no nest-building be-haviour would occur. However, Hughes and Duncan (1988) have reviewed this area and produced evidence that there are cases in which the performance of behaviour has motivationally-significant consequences that are not necessarily related to functional requirements. For example, hens perform nest-building sequences during pre-laying behaviour, even though the nest they created previously is still available (Hughes *et al.*, 1989) and dogs commonly turn round and round to flatten non-existent vegetation as a precursor to sleep.

Hughes and Duncan (1988) have developed a model of motivation to account for the occurrence of stereotyped, abnormal and repetitive behaviour often seen in barren or impoverished environments or when animals are highly motivated in situations where consummatory behaviour is difficult to carry out. In earlier models of motivation, such as those of Baxter (1983) and Wiepkema (1985), the functional consequences of behaviour have a negative feedback effect, acting both on motivation and through the animal's perception of external stimuli. To this basic model, Hughes and Duncan (1988) have added another feedback loop through which the performance of behaviour acts directly on motivation, the performance of appetitive behaviour acting positively and the performance of consummatory behaviour acting positively at first and then negatively. This means that for any behaviour pattern that is largely governed by internal factors, motivation will eventually increase above threshold. This will trigger appetitive behaviour, but in some artificial environments it may be impossible for the animal to proceed to the consummatory sequence. The appetitive behaviour will continue, sometimes in a fragmented form, and this will have the effect of raising motivation even further. The animal is then caught in a closed loop of performing apparently functionless behaviour from which it cannot escape until there is a change in its internal state. An example would be calves reared without their dams and bucket-fed. When these calves become hungry they have an increasing tendency to nuzzle and suck. In the absence of the dam, this behaviour is often directed towards the navels of penmates and the act of sucking increases the motivation to feed (and so to suck) even further. The provision of a bucket of milk may do little in the short term to reduce the navel sucking since the calves can consume the milk far more quickly than they could suck it from their dams and so the consummatory elements of feeding do not continue long enough to switch off the appetitive sucking. Eventually, of course, the functional consequences of consuming the milk will lead to satiation, but this may not occur for some time after the actual consumption.

Another example will demonstrate that, in any motivational system, it is difficult to predict (a) what will be the relative importance of attaining the goal compared with performing the behaviour and (b) what will be the constraining effect of a captive environment. Domestic hens prefer to lay in nests containing loose material which can be both moulded by their body and feet movements and manipulated with their beaks during nest building. Duncan and Kite (1989) investigated which of these two functions the hen regards as more important. They discovered that a substrate that allows the expression of moulding behaviour with body and feet movements appears to be crucial. Moreover, the substrate need not be mouldable; a pre-moulded nest which simulates the final moulded product is perfectly acceptable as long as it allows the hen to perform the moulding behaviour. The presence of loose material that can be manipulated with the beak does not seem to be regarded as essential by the nesting hen. Interestingly, if loose material is present, the hen will manipulate it with her beak. If it is absent, then many fewer manipulating motor patterns will be seen, but a small number may occur in vacuum. Rather like the turning behaviour of a dog before sleep, some manipulating with the beak appears to be pre-programmed into the hen's nest-building routine. Again it is important that the environment should allow the expression of this behaviour, but actual loose material to manipulate does not seem to be important to the nesting hen just as it does not seem important to the dog to have grass to flatten. These findings open up the opportunity for developing an artificial nest which can be incorporated into modern husbandry systems and which will improve the welfare of laying hens by allowing the full expression of nesting behaviour. The more general implications are that it may be necessary to design environments that not only allow the goal to be reached (say, the nest to be provided) but also allow the performance of the behaviour (say, the nest-building motor patterns to be performed). Moreover, the different motivational systems in each captive species will have to be examined in some detail to discover what are the important goals and behaviour patterns.

7.4 METHODS OF ASSESSING WELFARE

It is generally agreed that in any assessment of welfare, all the available evidence should be taken into account, including that of the animals' health, production, physiology and biochemistry as well as behaviour (Duncan, 1981, 1987). There are, however, problems associated with all these classes of evidence. For example, although injury and disease can be used as indicators of reduced welfare, their absence is not sufficient to prove well-being. For agricultural animals, production can also be a useful indicator, but only on an individual animal basis, which is often impossible to measure in practice. Again, good production is not sufficient to prove welfare. There are two problems associated with physiological and biochemical changes. One is the difficulty of taking the measurements without imposing further stress on the animals. The other is the predicament of calibration; how much of a change indicates reduced welfare? Because of the difficulties and limitations associated with all the other categories of evidence, there has been great interest in the use of behaviour as an index of welfare.

The advantages of using behaviour as an indicator of welfare include the fact that it can be observed without invasive techniques. In addition, it can be recorded without complicated equipment. Both these facts mean that it is an ideal tool for field use. Furthermore, disturbed behaviour may indicate more subtly and more quickly when the welfare of an animal is adversely affected.

There have been several different approaches which have used behaviour to assess welfare; some have been more successful than others. One method has been to study how animals behave in various states of suffering such as frustration, fear and pain. For example, it was claimed by the Brambell Committee that hens in battery cages will be frustrated because 'The normal productive pattern of mating, hatching and rearing young is prevented and the only reproductive urge permitted is mating. They cannot fly, scratch, perch or walk freely. Preening is difficult and dust-bathing impossible' (Command Paper 2836, 1965). This supposition can be, and has been, tested experimentally. Hens have been thwarted in many different ways while attempting to feed, behave sexually, nest, incubate eggs and brood chicks and their behaviour has been observed and catalogued. In these studies (Duncan, 1970), when frustration was severe, the symptoms were stereotyped back-and-forward pacing; when it was mild, there was an increase in an unusual type of preening which ethologists call 'displacement' preening, i.e. a behaviour pattern occurring out of context. If two or more hens were frustrated simultaneously then, in addition to the other responses, the dominant birds showed an increase in aggression towards the subordinates. When this catalogue was compared with the behaviour of hens in battery cages, it was found that displacement preening occurred quite frequently, suggesting that mild frustration is fairly common under commercial conditions. However, with one exception, the symptoms of severe frustration did not occur. The exception was that some hens showed symptoms of severe frustration in the pre-laying period when they did not seem able to regard the nest as a suitable nest site. So caging does not frustrate hens in all the ways suggested by the Brambell Committee, but it does thwart nesting, which is a possibility that they did not even consider.

Another state of suffering that has been widely investigated in domestic fowl is fear (Murphy, 1977, 1978; Murphy and Wood-Gush, 1978; Duncan, 1985; Jones 1987a,b). The stimuli that lead to a state of fear and the responses, both behavioural and physiological, made by the bird are now fairly well understood and this knowledge is being used to assess various husbandry procedures. For example, it has been shown that broiler chickens are *less* frightened when, prior to being transported to the slaughter house, they are caught by a well-designed 'harvesting machine' than when they are caught manually (Duncan *et al.*, 1986).

This method of stressing animals experimentally, recording their responses and comparing these responses with what occurs under commercial practice, has been extremely valuable in assessing the welfare of domestic fowl. Of course it does not tell us if animals, which are exhibiting symptoms of frustration or fear, are actually suffering. The initial assumption, that states of frustration, fear and pain will be aversive to the higher animals and will lead to a reduction in their welfare, depends upon arguments of analogy and homology between these animals and human beings — arguments which are powerful but not foolproof. However, the results from these studies lay the foundation for the next step, which is to investigate how the animals *feel* about being frustrated or frightened or in pain and this next step will be discussed later. Unfortunately, this method has not been used with other farm species or with laboratory or zoo species. It also has the disadvantage that some states of suffering may be missed. Frustration, fear and pain have been mentioned, but what of 'loneliness' or 'boredom'? Can animals suffer from those and, if so, can we recognize these states?

For the reasons given above, and also because there is a view that welfare should be more than simply the absence of states of suffering, other methods have been used to assess welfare. One of these has been to compare the behaviour of an animal in an environment which is assumed to be ideal with what occurs in the environment under investigation. In the case of wild species confined in zoos, this method may have some validity; generally speaking, species will be fairly well adapted to the natural environment in which they are found, and the behaviour shown there can be used as a 'benchmark'. However, in the case of our domesticated species, it is difficult to decide what is an 'ideal environment'. It is almost certainly not the natural environment in which their wild progenitors evolved. All of our domesticated farm and companion species have been artificially selected for thousands of years, our laboratory species for hundreds of generations, and this has resulted in them being substantially different from their ancestors.

With regard to farm species, perhaps 'traditional' husbandry systems would be near ideal since, until 40 years ago, they had changed only slowly, which meant that the species involved could also evolve under natural and artificial selection pressures to suit the system. However, 'traditional' systems often have obvious failings. For example, hill sheep in north-western Europe commonly suffer from malnutrition and exposure during severe winters to such an extent that many die. Another possibility is that feral populations of farm animals which are living and breeding in the wild may serve as a reference for behaviour in an ideal environment. However, the same criticisms can be raised as with animals under traditional systems; the animals are often severely stressed and exposed to hardships and deprivations which undoubtedly reduce their welfare. Another suggestion has been that an 'ideal environment' is one which elicits a rich repertoire of behaviour (Hughes, 1980). This can rapidly develop into a circular argument. Thus, if the 'ideal environment' has been chosen *because* it evokes a wide variety of behaviour, then it should hardly be surprising that when compared to this, other environments will appear 'poorer' in that they stimulate a more meagre range of behaviour.

Apart from the problem in deciding what is an 'ideal environment' for a domesticated species, this method provides results which are very difficult to interpret. It should be no surprise that animals behave differently in different environments and there is no guarantee that behaviour seen in more 'natural' environments indicates an adequate level of welfare. Differences in behaviour may simply demonstrate how adaptable animals are. Such differences *may* be suggestive of reduced welfare, but they will not become conclusive until they are corroborated by independent evidence. So the argument should be along the lines of the following: mice in laboratory cages show more stereotyped movements than mice in a terrarium and stereotyped movements have been shown independently to be indicative of frustration.

The interpretation of results from this method is particularly problematical when a behaviour pattern is missing in the test environment. It is too easy to say that the environment is 'preventing' the behaviour pattern in question without considering other reasons for its absence. The case of responses, such as anti-predator responses, that are elicited entirely by external stimuli, have already been discussed. So, some knowledge about the factors responsible for releasing and motivating the missing behaviour patterns is required before conclusions involving 'prevention' can be drawn.

A further problem is in interpreting behaviour patterns that occur in the test environment but in a vacuum, i.e. in the apparent absence of the external stimuli normally eliciting them. Sometimes the occurrence of vacuum activities seems to indicate a reduction in welfare. For example, Fraser (1975) has shown that sows which are kept in tether stalls without access to straw show vacuum manipulatory activities of the mouth and snout and these motor patterns often become very stereotyped and exaggerated. When straw is provided these activities are directed towards it and their frequency is greatly reduced, suggesting that there is a negative feedback effect when the behaviour is directed at an appropriate stimulus. On other occasions the occurrence of vacuum activities does not suggest reduced welfare and two examples have already been given. Nesting hens may perform vacuum nest-building motor patterns in the absence of loose nesting material and dogs, prior to sleeping, may perform vacuum turning and flattening movements in the absence of grass. Finally, there are vacuum activities about which it is difficult to draw conclusions on welfare. For example, hens in battery cages occasionally perform vacuum dust-bathing. Bouts of vacuum dust-bathing are much shorter and fragmented and occur less frequently than bouts of dust-bathing that take place when a dusty substrate is available. This suggests that, unlike nest-building motor patterns in nesting hens, bouts are not completely pre-programmed but are influenced by the environment (Vestergaard, 1980); dusty material appears to provide positive feedback, at least in the early part of bouts. However, if the performance of vacuum dust-bathing was not sufficient to reduce the internal causal factors (whatever they might be), one would predict that birds in cages would repeatedly start to dust-bathe, and this does not happen. This suggests that, in this case, the occasional performance of the vacuum activity may be sufficient to maintain welfare. In order to ensure welfare, however, another factor should be taken into consideration, namely that the performance of the vacuum activity should be non-damaging. In the case of vacuum dust-bathing in battery cages this is not always the case and hens frequently break both their feathers and their claws while doing it.

In trying to assess welfare, what we are interested in ultimately is what the animals subjectively 'feel' about what we do to them. It is possible that much of the behaviour that is suggestive of reduced welfare, such as frustration responses and fear responses, may be largely reflexive with little or no subjective feeling or awareness of what is going on. There has been much interest recently in the question of animal cognition, and there is general agreement that the higher animals (mammals and birds) are capable of at least the simpler cognitive processes (Griffin, 1976, 1984; Dawkins, 1980; Wood-Gush *et al.*, 1981; Walker, 1983). Although subjective feelings are not directly accessible to scientific investigation, there may be ways in which we can 'ask' animals indirectly what they think about the environment with which they are provided and the procedures to which they are subjected. Even a very crude measure of feelings, such as how positive or negative they are, would be extremely helpful in assessing welfare.

The simplest method is to give the animal a choice of various aspects of its environment and assume that it will express some of its feelings in its actions and choose in the best interests of its welfare. These techniques have been developed for studying preferences in domestic fowl and have proved useful in providing a first estimate of what they 'feel' about different aspects of their environment. In one of the first studies of this type, Hughes and Black (1973) showed that when hens were given a choice of four types of floor in a battery cage (Figure 7.2), the order of preference (as measured by the amount of time spent on each) was (1) wire-netting (which had been condemned by the Brambell Committee for not providing enough support), (2) conventional wire mesh but made of heavy gauge wire, (3) conventional wire mesh and (4) perforated sheet metal. In a later experiment, when given a choice between a wire mesh floor and a litter floor, hens

Figure 7.2 A hen being given the choice of two different types of flooring in a battery cage. The amount of time spent on each floor type is measured. The cage doors have been opened so that the flooring can be more easily seen.

did not demonstrate a clear-cut preference, but 88% of eggs were laid on the litter (Hughes, 1976b). In other experiments it was shown that hens preferred a large cage to a small one (Hughes, 1975a), an empty cage or one containing a small number of strange hens to one containing a large number of strange hens, and a cage containing familiar hens to one containing strangers (Hughes, 1977). Dawkins (1983a) has also been a strong advocate of this method. She found that when hens were given continuous access to a commercial battery cage and a large pen, they showed no preference. When they had to choose between a battery cage and an outside run, they showed a clear preference for the run, but choice was strongly influenced by prior experience. Giving birds access to food or companions in the battery cage did not markedly alter their preference for the outside run (Dawkins, 1976, 1977). Later she found that hens preferred a larger cage to a smaller cage and a cage with a turf floor to one with a wire floor (Dawkins, 1978). However, further studies showed that hens had a strong preference for litter floors over wire floors and that this could overcome their preference for large cage size. Hens showed a preference for litter when they could obtain access to it only by entering a small cage in which they could hardly turn round (Dawkins, 1981). Dawkins (1985) also investigated the preferences of hens for cages of different heights. Cages with less than 46cm height at the front and 37cm at the rear were actively avoided. In a parallel study, she showed that when hens were in cages of unrestricted height, 30% of all head movements took place above the common battery cage height of 38cm.

It is now recognized that simple preference tests have some short-comings and that there may be problems in interpreting the results from such tests. Duncan (1978) has listed three main areas where care is required:

1. The results only give information about the relative properties of two environments and not their absolute values. Therefore, as big a variety of choices as possible should be offered to guard against the animal choosing the better of two evils.
2. Minority choices are difficult to interpret; an animal which chooses to spend 10% of its time under one set of conditions may be making a positive choice which is as important for its welfare as the 90% choice.
3. An animal's short-term preference may not be in the best interests of its long-term welfare.

Thus a meat-type chicken kept for breeding purposes, if given free access to food, may eat too much and suffer from various painful orthopaedic diseases later in life (Duff and Hocking, 1986). Dawkins (1983b) has pointed out that a wild species in its natural environment will have proximate 'needs' which coincide with its ultimate 'needs'. However, that may not be true of a domesticated species in an artificial environment and we should not expect these animals to weigh up the long-term consequences of their decisions and make rational choices accordingly.

A development of the preference test is to use operant conditioning techniques to see how hard animals will work to obtain, or to avoid, some aspect of their environment. At first simple changes in the physical environment were used as rewards. For example, Baldwin and Meese (1977) showed that although pigs consistently preferred light to darkness, they were not prepared to work very hard to switch lights on. Similarly, sheep and calves worked to obtain much less light in motivational tests than they chose in preference tests (Baldwin and Start, 1980). Domestic fowl also preferred light to darkness and, when given the opportunity to switch light on and off, a laying strain chose to be illuminated for 77% and a broiler strain for 91% of the time. However, illuminated time fell to 20% when the lights went off automatically after three minutes and the birds had to switch them on again repeatedly (Savory and Duncan, 1982).

Later experiments have investigated more complicated phenomena. For example, van Rooijen (1983) showed that pigs would make a large number of operant responses in order to get from a concrete floor to an earthen floor. Operant techniques have also been used to measure the aversiveness of certain procedures. For example, Bailey *et al.* (1983) showed that when pigs were placed in a transport simulator which vibrated noisily, they would work in order to switch off the simulator for 30s. They worked harder when the vibrations were faster and when they had full stomachs. This study also showed that it was the vibrations and not the noise that was aversive to the pigs.

This approach is very promising but, once again, results require cautious interpretation particularly when an animal does not seem inclined to perform an operant response. Dawkins and Beardsley (1986) listed a number of reasons why an animal may not perform an operant response although it may be highly motivated. For example, the task that the animal has to perform (pecking a switch, depressing a lever, pushing a pad) may not always be appropriate for the reward it is receiving. Details of the spatial arrangements of the manipulanda and the temporal arrangements of the reinforcement schedule may not be appropriate for the response the animal is supposed to make. Furthermore, some behaviour may be less conditionable than other kinds.

In order to overcome the problems associated with operant conditioning methods, a technique has been developed in which the animal simply walks towards the reward (or away from the aversive stimulus or punishment) in a runway or simple maze. Its motivation to reach the reward (or to avoid the punishment) can be measured by placing various obstacles in the runway which it has to overcome to reach the reward (or to avoid the punishment) (Figure 7.3). An extension of this method is to see if animals will learn to walk through a maze following certain cues, such as coloured doorways, in order to reach a reward. Using this method it has been shown that domestic cocks will work quite hard to reach hens in order to court and copulate with them, but only when they can see the hens. When the hens are out of sight, cocks will not learn to push through a doorway in order to reach them. Hens will not push through a doorway in order to reach cocks whether they can see them or not (Duncan and Kite, 1987; Duncan and Hughes, 1988).

It is becoming increasingly obvious that welfare depends almost entirely on the cognitive needs of the animals involved (Duncan and Petherick, 1989). If this argument is carried to its ultimate conclusion, it means that even health is not a prerequisite for welfare; if an animal feels all right, then its welfare is all right. Of course there may be many other reasons for requiring good health in our animals, such as hygienic and productivity reasons. The findings from the experiments on sexual motivation in domestic fowl raise some interesting questions about the cognitive abilities of our captive animals. For example, are animals much more stimulus-bound than human beings who think about things that are absent from their current environment? In some cases the maxim may hold true that 'Out of sight is out of mind'. Also, do animals live much more in the present than human beings who think about past and future events? It is clear that we need to know much more about animals' cognitive abilities in order to assess welfare.

7.5 DESIGNING ETHOLOGICALLY-SOUND ENVIRONMENTS

7.5.1 Farm animals

Although there have been several attempts to design environments for our agricultural species which take account of their behaviour, few have set about this in a systematic way. An exception has been the work of Stolba and Wood-Gush in developing the 'family pen system' for housing pigs (Stolba, 1981; Stolba and Wood-Gush, 1981). Their

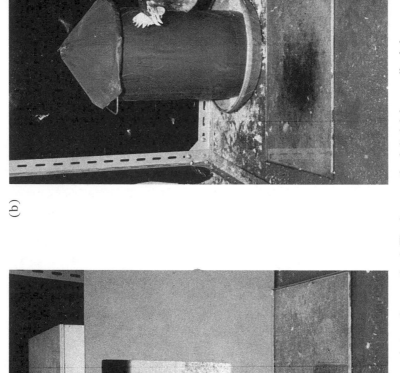

(b)

(a)

Figure 7.3 Measuring motivation by the obstruction method. The hen on the left (a) has walked down a runway towards a familiar red nestbox and through a shallow footbath (which hens find aversive) in order to enter the nest and lay an egg. Nesting motivation can thus be compared with hunger by seeing how hungry a hen will be before it crosses the foot-bath in order to reach food in a familiar blue feeder (b).

approach was to study the function and development of the behaviour of a small population of pigs living with minimal interference on a partially wooded hillside in Scotland. They were able to identify certain key features in the environment which consistently seemed to guide the performance of frequent or regular sequences of behaviour. These key environmental features either stimulated or 'released' important behavioural sequences or provided a source of reference so that the behaviour was correctly orientated. The next stage in the study involved confining the pigs in a substantially smaller area on the hillside but with all the key features present. Then certain of the key features were removed in a systematic way. The effects of both of these operations on the patterning and orientating of behavioural sequences was recorded and this allowed the relative importance of space and environmental richness to be assessed. In a similar way, key items in the social environment were identified. A housing system was then gradually developed which required neither a large area nor expensive buildings and which incorporated the essential key features as defined areas with 'furniture'. The basic unit of the system was a series of interconnecting pens to take four sows and their followers (see Figure 6.7, Chapter 6). The pens were partially roofed and partially open to simulate a forest border where many activities took place on the hillside (Figure 7.4). Each sow was provided with a nest site, sheltered against the elements but with a good view out of the pen through the front gate. Space was allowed for a nest 2–3m in diameter, which was the size of nest built on the hillside. Straw for building the nest was provided in a rack some distance away. The sequence of pulling straw from the rack, gathering it in the mouth, carrying it to the nest site and depositing it with nodding movements was thus possible. Although the journeys were shorter and the nesting material more homogeneous than on the hillside, the organization of this behaviour was almost identical and it was appropriately orientated. The same was true of nest-building behaviour. A corridor was provided for defaecating 4.5–11m away from the nest. It had been observed that pigs walked this distance from the nest on the hillside before defaecating on paths between bushes. Also provided were activity areas containing straw, areas containing peat and bark for rooting and wallowing, horizontal bars which the pigs could lever with their snouts, sufficient space for all pigs to feed simultaneously and refuges for any pigs coming under social pressure. Using this method, Stolba and Wood-Gush have managed to design a pig housing system which takes much more account of the animals' behaviour than any modern husbandry system. Although no absolute measure is available, there can be little doubt that these pigs enjoy a high level of welfare; all the symptoms

(a)

(b)

Figure 7.4 Many activities of a population of pigs living in a natural habitat in Scotland took place at the forest border (a). The family pens developed by Stolba and Wood-Gush (1981) were partially roofed and partially open in order to simulate the forest border (b).

which are suggestive of reduced welfare are absent. For example, all types of stereotypies are absent, vices such as tail-biting and canni-balism are absent, aggression is at a low level, much more time is spent in manipulating and exploring objects than other pigs and fear responses are not exaggerated. Moreover, the family pen system replaces dry sow stalls, farrowing crates and the modern fattening house, the three areas of pig production that have attracted the most severe criticisms on welfare grounds. In addition, the system is promising commercially.

Attempts to design ethologically-sound environments for other classes of farm livestock have been much less methodical and, perhaps as a result of this, less successful. Of course, in some cases there have been insuperable difficulties. A crucial feature of the system devised by Stolba and Wood-Gush is that it is a *family* system which takes account of the pigs' social structure. It is difficult to envisage how, for example, a veal calf system could be designed along 'family' lines when the calves are a by-product of the dairy system and their dams are required to produce milk for human consumption. Of course it may be possible to design an artificial mother which fulfils some of the functions of the natural mother as has been done for piglets (Jeppeson, 1981; Lewis *et al.*, 1982) but it is unlikely that this would fully compensate for the lack of the mother–offspring bond. Nevertheless, husbandry systems for veal calves have been developed which take account of some of the calves' behaviour and which have welfare benefits over the conventional crate system. For example, Webster and Saville (1981) showed that keeping calves on the 'Quan-tock System' (Paxman, 1981), i.e. in groups in a straw yard and with their liquid diet supplied through teats, resulted in three major welfare benefits compared with conventional crates. These were social contact, roughage to eat and ability to adopt a normal sleeping posture.

There have been many attempts to design alternative husbandry systems to the battery cage for laying hens. These attempts have been beset with similar constraints to those which have restricted the development of veal calf housing. The drive for efficiency in the egg industry has meant that the presence of 'unproductive' males would not be countenanced, although there is evidence that the presence of males might have a beneficial effect on the social organization of a group of hens (Bhagwat and Craig, 1979). Also the enormous improve-ments that have been made in poultry hygiene in recent years have been accomplished by employing the policy of having only one age group of birds 'all in' and 'all out' of a poultry house at one time. The alternative systems to battery cages have thus concentrated on improving the physical environment and have avoided interfering with the social environment.

One approach has been to modify the battery cage in various ways. The simplest modifications to improve welfare are design changes to reduce the risk of birds trapped and injured and Tauson (1980) has made much progress in this direction. A more radical approach has been to enrich the battery cage by enlarging it, particularly in the vertical dimension, providing perches and a nesting site and, in some models, providing a dust bath. This type of enriched cage has become known by the generic name of 'get-away cage' (Figure 7.5). It was developed in the UK independently by Bareham (1976) and Elson (1976) and later in the Netherlands, West Germany and Switzerland (Wegner *et al.*, 1981; Brantas *et al.*, 1978). There is no doubt that the get-away cage caters for more of the hen's needs than a conventional cage; it takes account of the bird's preference for a larger cage and litter substrate and avoids the problems of pre-laying frustration. However, there has been a price to pay. Because of the expense and the design difficulties involved in providing the additional facilities for a small group of birds, the group size in get-away cages has been increased from the usual 3–5 to 10–25. This is greater than the optimum group size for laying hens (Hughes, 1975b) and probably means that social friction is increased. Other difficulties have included feather soiling resulting from the vertical layout of these cages. In addition, there have been production problems such as dirty and cracked eggs and a higher food intake than in conventional cages (Elson, 1981). It seems that the ideal get-away cage has not yet been designed but development is continuing.

Another approach has been to modify the deep-litter system. This system, which depends on the activity of micro-organisms in the litter to kill off pathogens, has never worked well in north-west Europe, probably because of the cool damp winters; if the litter stops 'working' and becomes cold and wet, disease outbreaks invariably follow. Alternative systems have tried to overcome this problem by increasing the bird numbers (and therefore the heat production) within a house by making more use of vertical space. There have been many variations on this theme, such as the 'perchery' which incorporates frames with perches (McLean *et al.*, 1986), various types of 'aviary' with slatted or wire platforms (Hill, 1981a,b), and 'tiered wire floors' (Ehlhardt, 1985). All of these systems allow stocking density to be increased and provide various facilities at different vertical levels. While they undoubtedly allow the birds more freedom to express a greater range of behaviour than do cages, there have been associated husbandry problems some of which may actually reduce welfare. For example,

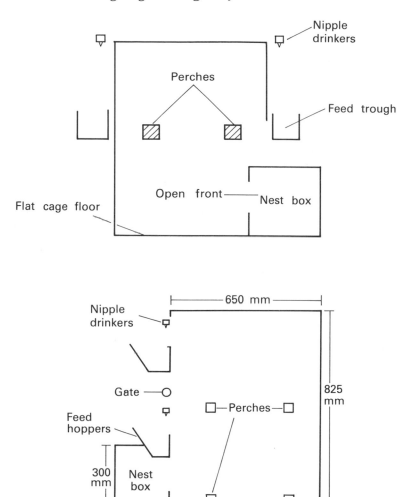

Figure 7.5 Cross sections through two typical get-away cages (from Elson, 1981).

ammonia levels have been high and social friction, as measured by comb damage and loss of head feathers, has been higher than in cages (Hill, 1981a,b). There have also been management problems such as high food consumption and a high incidence of dirty and cracked eggs which have made it difficult for these alternative systems to compete commercially with battery cages. It is to be hoped that research and development will continue until a husbandry system is found which truly caters for the birds' welfare and which also can match the battery cage commercially.

7.5.2 Laboratory animals

During the last 40 years, since the appearance of the first edition of the Universities Federation For Animal Welfare (UFAW) Handbook on the Care and Management of Laboratory Animals, there have been enormous improvements in the design and standards of housing and caging for laboratory animals. Most of these improvements have been concerned with the production of healthy stock and hygienic conditions. Where necessary, some species of laboratory animals have been produced by caesarian section under sterile conditions, so that their microflora and disease status are known. Such animals are maintained behind barriers and are available for microbiological and immunological studies. Air conditioning with humidity control and a closely defined temperature range are nowadays taken for granted and every effort is made to maintain animals within their physiological optimum range of environmental conditions (Clough, 1984). The production of genetically inbred strains of rodents has resulted in decreased variability and hence a reduction in the number of animals used. Carefully balanced standardized diets are available as are uncontaminated material for bedding and substrates. Cages present smooth impenetrable surfaces which can be easily sterilized (Poole, 1987).

A number of changes in design of animal houses and their equipment have been initiated in order to save labour. However, such developments may not always promote the welfare of the animals because the reduction in contact between staff and animals may mean that there is less awareness of the animals' needs.

The high cost of specialized housing necessitates careful consideration of economics, and, for this reason, there has been an increase in caging density in the animal house and often a reduction in living space for the animals. Various guidelines have been produced to define the minimum spatial allowances for different laboratory animals (ILAR, 1980; Council of Europe, 1986; Royal Society/UFAW, 1987). However, these are based on current practice rather than the scientific assessment of the physiological and psychological needs of the different species.

A large number of vertebrates are used in laboratories, the majority of which are rodents but with a range of mammalian species from the grey opossum (*Monodelphis domestica*) to the chimpanzee (*Pan troglodytes*). In addition, various species of birds, fish, reptiles and amphibia and a small number of invertebrates are also used (Poole, 1987).

In the past, most ethologists have tended to study wild species in natural or semi-natural environments. Applied ethologists have concentrated mainly on farm animals and their well-being. For these reasons, there is a serious lack of knowledge as to how the average laboratory animal behaves in its cage and therefore what its needs may be. Veterinary supervision ensures that the animal's physical condition is monitored, but few veterinary courses include training in behavioural methodology, so that the animal's behavioural requirements may not be recognized. Currently, housing provides all the metabolic needs of the laboratory animal and protects it from disease and parasites but whether or not it caters for the animal's behavioural needs is a matter of chance, since these have seldom been identified.

Common laboratory rodents can be regarded as domesticated, since they are docile and readily handled, show a minimum of neophobia, are generally unreactive and exhibit little fear of, and a much reduced flight distance to, man. However, even though laboratory animals are adapted to a captive environment, much of their natural repertoire of behaviour is retained. For example, laboratory mice build nests, form territories, protect their young and males are hostile to strange males. The patterns of social behaviour observed do not differ from those of their wild counterparts. In spite of this, in the laboratory, mice and rats are usually held in relatively featureless cages with only a sawdust substrate and *ad lib* food and water. The most serious threat to their well-being probably results from high levels of ambient ultrasound (Sales *et al.*, 1989) and, in male mice, aggression from individuals in the same cage. More research is needed to determine which types of environmental ultrasound should be eliminated (Sales and Wilson, 1986) and to determine optimum social housing for male stock mice (Goldsmith *et al.*, 1976; Brain, 1989). The spatial requirements of most laboratory rodents are not known, nor is it clear whether they benefit from any kind of environmental enrichment within the cage (Wallace, 1984).

The most serious problems in designing ethologically sound housing systems are found in maintaining old world monkeys and apes in captivity (Goosen, 1988; Poole, 1988a; Harris, 1988). These animals are not domesticated and have complex requirements for stimuli and companionship and yet are often housed singly in cages which do not allow them to climb, run or jump let alone form intricate social relationships (see also Chapter 3). In many instances they show abnormal behaviour such as locomotor stereotypies and self-directed aggression (Goosen, 1989; Harris, 1988). Some laboratories have devised excellent housing for these species (Jaeckel, 1989), but in most cases suboptimal conditions are provided (Figure 7.6). One of the greatest dilemmas is how to give the animals adequate social contact when there is a need for separation as part of the experimental paradigm or for quarantine reasons. It must be emphasized that very few experiments actually require that the animals should be physically separated. For example, animals given the same dose of a substance or infected with the same disease organism should always be housed socially. However, in studies in which metabolic rates or feeding rates are required, there is little alternative to keeping animals isolated. While it is clearly unsatisfactory from a welfare point of view for these animals to be deprived of tactile contact with conspecifics, it is not known the extent to which such deprivation may invalidate the results of these experiments. It is becoming apparent that certain experimental manipulations, such as early separation of infant primates from their mothers which can lead to behavioural problems, may also have effects on both endocrine and immune systems (Martin, 1989).

There is an increasing awareness of the importance of meeting the needs of captive primates and a number of strategies have proved beneficial. Chamove *et al.* (1982) showed that a wood chip substrate with food included in it increased foraging and cut down aggression in seven species of primate when socially housed. Harris (1988) found that such a substrate reduced the levels of locomotor stereotypies in singly-housed long tailed macaques (*Macaca fascicularis*). A number of workers have found that electronic keyboards, puzzle-feeders and novel objects will improve the captive situation for monkeys (Markowitz, 1982). Reinhart *et al.* (1988) have shown that long-term singly-housed female rhesus monkeys can be successfully housed socially with a compatible female if their behaviour is closely monitored; only 16% proved incompatible. The problem of socially housing male macaques and baboons is more difficult to solve. However, the formation of same-sex peer groups of young individuals may allow males to be housed together subsequently, providing that they are out of contact with females. Undoubtedly, from the aspect of welfare, the most suitable primate species to keep in captivity is the common marmoset (*Callithrix jacchus*). This monkey can be maintained in spacious conditions in family groups of up to ten members. It adapts well to captivity and it is easy to provide cage furniture to create a complex environment.

A particularly important aspect of the social behaviour of primates is their prolonged mother–infant bond (see Chapter 6 for other species). It has been traditional for laboratory-bred primates to be

Figure 7.6 A long tailed macaque housed singly in a typically barren laboratory cage.

weaned and separated from their mother at an early age (3–6 months as opposed to 12–18 months in the wild). The object of such early weaning is to increase productivity by terminating the mother's lactation to bring her into oestrus sooner. This strategy is both stressful to the infant and counter productive; it is unnecessary in callitrichids which have a post-partum oestrus and in rhesus monkeys which are, in any case, seasonal breeders. In *Macaca fascicularis*, although early weaning may lead to some increase in productivity, there is evidence from other macaque species that the resulting young may show high levels of abnormal behaviour (Goosen, 1988) and this should be borne in mind.

7.5.3 Zoo animals

The aim of the zoo is to display animals to the public against a naturalistic background where they can behave in as complex a manner as possible. This represents an ideal and, in the case of park deer, antelope and rhinoceros, a large grassy paddock may suffice to provide a good environment and the animals will still be visible to the public. There are greater challenges, however, when the animal is small, nocturnal or cryptic in habits or appearance. For example, if monkeys are provided with a large amount of vegetation, they will be much less visible to the public and, as a result, form a much less attractive exhibit. In many species housing must represent a compromise between a naturalistic setting and a high degree of visibility to the viewer. There is now a much greater awareness on the part of the public with regard to the welfare of zoo animals. The abnormal behaviour induced in many species by confinement in a bare and featureless environment are apparent to all but the least observant. As a result the better zoos make an effort to provide their stock with as good an environment as they can.

It is the more intelligent omnivores and carnivores which require the most complex environments and these are precisely the species which require the greatest security in caging. To provide for the welfare of these species is the greatest challenge. Two kinds of solution to the problems of what appear to be apathy, boredom and the development of abnormal behaviour have been attempted and these can be broadly defined as naturalistic and mechanistic solutions. An example of the former is the provision of an artificial termite mound for great apes; the artificial mound contains sweet food which the animal can reach by poking a tool, such as a twig, through a hole in the mound to obtain the reward (Nash, 1982; McEwen, 1986). There are now increasing numbers of exhibits in zoos which use inexpensive methods of enriching the environment by simulating nature. Good examples of these are the recipients of the UFAW Zoo Animal Welfare Awards; Drusilla's Zoo Park created a naturalistic setting for beavers, which includes an underground lodge and facilities for damming a stream (Figure 7.7). Belfast Zoo designed an enclosure for gentoo penguins (*Pygoscelis papua*) which incorporates a filtered swimming

Figure 7.7 A view of the beaver exhibit at Drusilla's Zoo Park showing branches from which the animals have constructed a dam. There is a large pool and grassy area with a subterranean lodge with underwater entrance.

pool, self-draining pebble beach and a grassy area. This simulates the birds' natural environment in the Falkland Islands and enables them to show a wide repertoire of behaviour including nest building and rearing young (Figure 7.8). At the London Zoo, the dwarf mongoose enclosure includes simulated termite mounds. There are dispensers for crickets and mealworms from which the prey emerges at random from holes in the container. The enclosure also features sun and cloud effects to stimulate sun bathing and a sandy substrate is provided so that food, such as whole eggs, can be scattered and hidden. The dwarf mongooses, under these circumstances, carry out a wide range of natural behaviour. This exhibit, particularly, shows that inexpensive low technology solutions can provide behavioural enrichment for captive animals. The behaviour of the mongooses has been carefully monitored and compared with that prior to environmental enrichment and the welfare improvement has been fully vindicated (Figure 7.9).

A reaction timer which delivered a reward to a mandrill on pressing a button sufficiently soon after a signal light came on, provides an example of a mechanistic solution to environmental enrichment. The timer was programmed in such a way that the animal could play

Figure 7.8 Gentoo penguins with chick born in Belfast Zoo's penguin enclosure. The picture shows the pebble beach with grass tussocks.

against either a member of the public or a computer (Markowitz, 1982). Growing vines in primate cages represents a compromise between the mechanistic and the natural. Gibbons which swung on these vines obtained a food reward, thus providing a foraging experience and at the same time encouraging activity on the part of the animal.

All these devices represent praiseworthy attempts to improve the environment for different species. However, it is most important that the efficacy of different methods should be scientifically assessed. What is clear is that simple procedures such as scattering food unpredictably in the animal's environment may be both cheap and effective. There is, however, considerable controversy as to the justifiability and effectiveness of different means of environmental enrichment in the zoo (Forthman Quick, 1984).

In considering zoos, it is clear that the greater available space, as compared with the laboratory, and the small numbers of each species kept, compared with agricultural practice, make environmental enrich-

Figure 7.9 Dwarf mongooses in the London Zoo Clore Pavilion. The large log in the foreground contains live crickets which emerge at intervals from a hole.

ment much easier. However, laboratory animals are maintained for the serious purpose of biomedical research, whereas some would regard the keeping of animals in zoos as both unnecessary and frivolous. Certainly, zoos need to justify keeping animals in captivity and one must always bear in mind the three main functions of a zoo which are education, conservation and research (UFAW, 1988). It should also be remembered that animals in zoos, whilst they are confined, are protected from environmental hazards encountered in the wild such as predators, diseases and competition. The aim of the zoo should be to provide every animal with an environment in which it can show as wide a range of natural behaviour as is practicable and desirable.

If serious efforts are to be made to improve the captive environment in laboratories and zoos, the animals' behaviour must be monitored to discover how their time is spent, and these data should then be used to assess their quality of life. Aspects of behaviour clearly indicative of poor welfare are abnormal patterns such as repetitive stereotypies, self-mutilation and a very restricted behavioural repertoire. Definitions and descriptions of abnormal behaviour in primates have been given by Erwin and Deni (1979) and Poole (1988b). The same approach could be taken as has been carried out with domestic fowl (see section 7.4) in which the aim is to recognize indicators of states of suffering such as frustration and fear. The behaviour observed in a particular captive environment can then be examined for these indicators and compared with that observed in the wild or in complex captive conditions. If welfare problems are detected, it is not only important that the environment is enriched, but also that behaviour is monitored during and after the changes to the environment. This will ensure that any reduction in abnormal behaviour and any changes in the incidence and frequency of natural behaviour will be recorded and so welfare under the new environmental conditions will be properly assessed.

It is clear that there are few instances where the environmental/behavioural needs of the species kept in zoos are fully understood and this is especially true of animals whose sensory systems are very different from our own, for example rodents. There is a pressing need for more research on animal welfare and there is evidence that scientists are becoming increasingly aware of the importance of this field (UFAW, 1989). Unfortunately, this area of research does not attract mainstream scientific funding because it is not seen to be either basic academic or applied research. There is no doubt, however, that animals which are in impoverished environments or stressed make poor subjects for both public exhibition and research projects. Some organizations in the UK, such as UFAW and the Association for the Study of Animal Behaviour, provide small-scale funding for behavioural animal welfare research but this remains a seriously underfunded field.

7.6   OTHER SOLUTIONS

This chapter has been concerned with how ethological studies can promote the welfare of captive animals. It has defined what is meant by 'welfare', shown how it may be assessed and given examples of how captive environments may be modified in order to improve welfare. There is, however, another possible route that may be taken in order to protect welfare and that is by changing the animals genetically. This route is of no interest to zoos since their aim is to keep representatives of wild species in captivity. However, there are many possibilities for agricultural animals and for animals kept for research purposes. In its simplest form this process is the traditional one of domestication whereby animals are selected for breeding that are more tractable and show less flight distance to man. Oldfield-Box (1981) has stated that domestication also involves making domesticants more suitable to specific environmental conditions which relate to nutrition, housing and climate. Beilharz (1982) has also emphasized this evolutionary adaptation to the environment which accompanies or is an integral part of domestication. He says that there is no reason why domestic species should not be able to change further and so be better adapted to intensive conditions. He admits that when animals are first placed in environments to which they are not adapted they will be stressed and show inappropriate behaviour which will result in injury and lowered production, in other words, reduced welfare. After a period of adaptation, genetic shifts will result in a new strain of domestic animal that is adjusted to the new conditions. Apart from the question of whether or not it is ethically acceptable knowingly to subject animals to a period of adaptation and genetic change, this philosophy is dangerous for three reasons. First, although behaviour may evolve relatively quickly, it is still going to take many generations and in the past 20 years husbandry systems have come and gone more quickly than this. Second, animals cannot adapt genetically to a particular environment when the breeding stock are not kept in the same environments as the commercial stock, a common agricultural practice. Third, the artificial selection pressure for certain production characteristics may be so strong that it overrides the 'natural' selection for animals that are adapted.

It is obvious then that we cannot rely on the traditional process of domestication to adapt animals to the captive environments that we provide for them, but it may be possible to carry out selection procedures for or against specific traits, such as fearfulness (Faure, 1980), that will lead to improved adaptation. Few people would have qualms about, say, selecting strains of cattle that were less nervous and so better suited to milking parlours. It is a small step from there, to selecting for animals that are better adapted by being 'deficient' in some aspect of their behaviour. For example, Mills and Wood-Gush (1982) have shown that it is possible to select hens that do not show symptoms of pre-laying frustration but sit quietly in what many people would consider as the 'inadequate' environment of a battery cage.

Also, it might be possible to select birds that have a much reduced tendency to dust-bathe and so are better adapted to cages (Gerken and Petersen, 1987). The ethics of such procedures are questionable, especially as the logical conclusion would be to select for, or genetically engineer, animals that are much less aware of their environment and are, in effect, little more than vegetables. Why not move in the other direction of cell and tissue culture for the production of edible protein and a medium for many medical and veterinary experiments? Such a move would be ethically acceptable to most people and would do much to reduce animal suffering.

REFERENCES

Bailey, K.J., Stephens, D.B., Ingram, D.L. and Sharman, D.F. (1983) The use of a preference test in studies of behavioural responses of pigs to vibration and noise. *Appl. Anim. Ethol.*, **11**, 197.

Baldwin, B.A. and Meese, G.B. (1977) Sensory reinforcement and illumination preference in the domesticated pig. *Anim. Behav.*, **25**, 497–507.

Baldwin, B.A. and Start, I.B. (1980) Studies on illumination preference and sensory reinforcement in sheep and calves. *Appl. Anim. Ethol.*, **6**, 389–90.

Bareham, J.R. (1976) A comparison of the behaviour and production of laying hens in experimental and conventional battery cages. *Appl. Anim. Ethol.*, **2**, 291–303.

Baxter, M.R. (1983) Ethology in environmental design for animal production. *Appl. Anim. Ethol.*, **9**, 207–20.

Beilharz, R.G. (1982) Genetic adaptation in relation to animal welfare. *Int. J. Stud. Anim. Prob.*, **3**, 117–24.

Bhagwat, A.L. and Craig, J.V. (1979) Effects of male presence on agonistic behaviour and productivity of White Leghorn hens. *Appl. Anim. Ethol.*, **5**, 267–82.

Brain, P.F. (1989) Social stress in laboratory mouse colonies, in *Laboratory Animal Welfare Research*, pp. 49–61, Proceedings of a UFAW Symposium. UFAW, Potters Bar.

Brantas, G.C., De Vos-Reesink, K. and Wennrich, G. (1978) Ethologische Beobachtungen bei Legehuhnern in Get-away-Kafigen. *Arch. Geflugelk.*, **42**, 129–32.

Carpenter, E. (1980) *Animals and Ethics*, Watkins, London.

Chamove, A.S., Anderson, J.R., Morgan-Jones, S.C. and Jones, S.P. (1982) Deep woodchip litter: hygiene, feeding and behavioral enhancement in eight primate species. *Int. J. Stud. Anim. Prob.*, **3**, 308–18.

Clough, G. (1984) Environment factors in relation to the comfort and well-being of laboratory rats and mice, in *Standards in Laboratory Animal Management*, pp. 7–24, Proceedings of a LASA/UFAW Symposium. UFAW, Potters Bar.

Command Paper 2836, 1965. Report of the Technical Committee to Enquire into the Welfare of Animals kept under Intensive Livestock Husbandry Systems, H.M.S.O. London.

Council of Europe (1986) European Convention for the Protection of Vertebrate Animals used for Experimental and other Scientific Purposes, Council of Europe, Strasbourg.

Dawkins, M.S. (1976) Towards an objective method of assessing welfare in domestic fowl. *Appl. Anim. Ethol.*, **2**, 245–54.

Dawkins, M.S. (1977) Do hens suffer in battery cages? Environmental preferences and welfare. *Anim. Behav.*, **25**, 1034–46.

Dawkins, M.S. (1978) Welfare and the structure of a battery cage: size and cage floor preferences in domestic hens. *Br. Vet. J.*, **134**, 469–75.

Dawkins, M.S. (1980) *Animal Suffering*, Chapman and Hall, London.

Dawkins, M.S. (1981) Priorities in the cage size and flooring preferences of domestic hens. *Br. Poult. Sci.*, **22**, 255–63.

Dawkins, M.S. (1983a) The current status of preference tests in the assessment of animal welfare, in *Farm Animal Housing and Welfare* (eds S.H. Baxter, M.R. Baxter and J.A.C. MacCormack). pp. 20–6, Martinus Nijhoff, The Hague.

Dawkins, M.S. (1983b) Battery hens name their price: consumer demand theory and the measurement of ethological 'needs'. *Anim. Behav.*, **31**, 1195–205.

Dawkins, M.S. (1985) Cage height preference and use in battery-kept hens. *Vet. Rec.*, **116**, 345–7.

Dawkins, M.S. and Beardsley, T.M. (1986) Reinforcing properties of access to litter in hens. *Appl. Anim. Behav. Sci.*, **15**, 351–64.

Driscoll, J.W. and Bateson, P. (1988) Animals in behavioural research. *Anim. Behav.*, **36**, 1569–74.

Duff, S.R.I. and Hocking, P.M. (1986) Chronic orthopaedic disease in adult male broiler breeding fowls. *Res. Vet. Sco.*, **41**, 340–8.

Duff, S.R.I., Hocking, P.M. and Field, R.K. (1987) The gross morphology of skeletal disease in adult male breeding turkeys. *Avian Path.*, **16**, 635–51.

Duncan, I.J.H. (1970) Frustration in the fowl, in *Aspects of Poultry Behaviour* (eds B.M. Freeman and R.F. Gordon), pp. 15–31, British Poultry Science Ltd, Edinburgh.

Duncan, I.J.H. (1978) The interpretation of preference tests in animal behaviour. *Appl. Anim. Ethol.*, **4**, 197–200.

Duncan, I.J.H. (1981) Animal rights — animal welfare: a scientist's assessment. *Poult. Sci.*, **60**, 489–99.

Duncan, I.J.H. (1985) How do fearful birds respond? in *Second European Symposium on Poultry Welfare* (ed. R.-M. Wegner), pp. 96–106, W.P.S.A. Celle.

Duncan, I.J.H. (1987) The welfare of farm animals: an ethological approach. *Sci. Prog., Oxf.*, **71**, 305–15.

Duncan, I.J.H. and Dawkins, M. (1983) The problem of assessing 'well-being' and 'suffering' in farm animals, in *Indicators Relevant to Animal Welfare* (ed. D. Smidt), pp. 13–24, Martinus Nijhoff, The Hague.

Duncan, I.J.H. and Hughes, B.O. (1988) Can the welfare needs of poultry be measured? in *Science and the Poultry Industry* (ed. J.Y. Hardcastle), pp. 24–5, A.F.R.C., London.

Duncan, I.J.H. and Kite, V.G. (1987) Some investigations into motivation in the domestic fowl. *Appl. Anim. Behav. Sci.*, **18**, 337–8.

Duncan, I.J.H. and Kite, V.G. (1989) Nest site selection and nest-building behaviour in domestic fowl. *Anim. Behav.*, **37**, 215–31.

Duncan, I.J.H. and Petherick, J.C. (1989) Cognition: the implications for animal welfare. *Appl. Anim. Behav. Sci.*, **24**.

Duncan, I.J.H., Slee, G., Kettlewell, P. *et al.* (1986) Comparison of the stressfulness of harvesting broiler chickens by machine and by hand. *Br. Poult. Sci.*, **27**, 109–14.

Ehlhardt, D.A. (1985) Tiered floors an acceptable compromise. *Misset Intern. Poult.*, Jan. 1985, 58–61.

Elson, A. (1976) New ideas on laying cage design — the get-away cage. *Proc. V Europ. Poult. Conf., Malta*, pp. 1030–41.

Elson, A. (1981) Modified cages for layers, in *Alternatives to Intensive Husbandry Systems*, pp. 47–50, U.F.A.W., Potters Bar.

Erwin, J. and Deni, R. (1979) Strangers in a strange land: abnormal behaviors or abnormal environment? in *Captivity and Behaviour* (eds J. Erwin, T.L. Maple and G. Mitchell), pp. 1–28, Van Nostrand Reinhold, New York.

Faure, J.M. (1980) To adapt the environment to the bird or the bird to the environment? in *The Laying Hen and its Environment* (ed. R. Moss), pp. 19–42, Martinus Nijhoff, The Hague.

Flock, D.K. (1986) EEC poultry production, past, present and future. *Misset Intern. Poult.*, Oct. 1986, 6–9.

Forthman Quick, D.L. (1984) An integrative approach to environmental engineering in zoos. *Zoo Biol.*, **3**, 54–77.

Fraser, D. (1975) The effect of straw on the behaviour of sows in tether stalls. *Anim. Prod.*, **21**, 59–68.

Gerken, M. and Petersen, J. (1987) Bidirectional selection for dust-bathing activity in Japanese quail (*Coturnix coturnix japonica*). *Br. Poult. Sci.*, **28**, 23–37.

Gerrits, A.R., De Koning, K. and Migchels, A. (1985) Catching broilers. *Misset Intern. Poult.*, July, 1985, 20–3.

Goldsmith, J.F., Brain, P.F. and Benton, D. (1976) Effects of age at differential and the durations of individual housing/ grouping on intermale fighting behaviour and adrenocortical activity in 'TO' strain mice. *Aggress. Behav.*, **2**, 307–23.

Goosen, C. (1988) Developing housing facilities for rhesus monkeys: prevention of abnormal behaviour, in *New Developments in Biosciences: Their Implications for Laboratory Animal Science* (eds A.C. Beynen and H.A. Solleveld), pp. 67–70. Martinus Nijhoff, Dordrecht.

Goosen, C. (1989) Influence of age of weaning on the behaviour of rhesus monkeys, in *Laboratory Animal Welfare Research, Primates* pp. 17–22, Proceedings of a UFAW Symposium, UFAW, Potters Bar.

Grandin, T. (1983) Welfare requirements of handling facilities, in *Farm Animal Housing and Welfare* (eds S.H. Baxter, M.R. Baxter and J.A.D. MacCormack), pp. 137–49, Martinus Nijhoff, The Hague.

Griffin, D.R. (1976) *The Question of Animal Awareness*, Rockefeller University Press, New York.

Griffin, D.R. (1984) *Animal Thinking*, Harvard University Press, Cambridge, Mass.

Harris, D. (1988) *Welfare and Housing of Laboratory Primates*, UFAW Research Project No. 1, UFAW, Potters Bar.

Hediger, H. (1950) *Wild Animals in Captivity*, Butterworth, London.

Hill, J.A. (1981a). The aviary system, in *First European Symposium on Poultry Welfare* (ed. L.Y. Sorensen), pp. 115–23, W.P.S.A., Copenhagen.

Hill, J.A. (1981b) Aviary systems for layers, in *Alternatives to Intensive Husbandry Systems*, pp. 40–6, UFAW, Potters Bar.

Hughes, B.O. (1975a) Spatial preference in the domestic hen. *Br. Vet. J.*, **131**, 560–4.

Hughes, B.O. (1975b) The concept of an optimum stocking density and its selection for egg production, in *Economic Factors Affecting Egg Production* (eds B.M. Freeman and K.N. Boorman), pp. 271–98, British Poultry Science Ltd, Edinburgh.

Hughes, B.O. (1976a) Behaviour as an index of welfare. *Proc. V Europ. Poult. Conf., Malta*, pp. 1005–18.

Hughes, B.O. (1976b) Preference decisions of domestic hens for wire or litter floors. *Appl. Anim. Ethol.*, **2**, 155–65.

Hughes, B.O. (1977) Selection of group size by individual laying hens. *Br. Poult. Sci.*, **18**, 9–18.

Hughes, B.O. (1980) Behaviour of the hen in different environments. *Anim. Reg. Stud.*, **3**, 65–71.

Hughes, B.O. and Black, A.J. (1973) The preference of domestic hens for different types of battery cage floor. *Br. Poult. Sci.*, **14**, 615–19.

Hughes, B.O. and Duncan, I.J.H. (1988) The notion of ethological 'need', models of motivation and animal welfare. *Anim. Behav.*, **36**, 1696–1707.

Hughes, B.O., Duncan, I.J.H. and Brown, M.F. (1989) The performance of nest building by domestic hens: is it more important than the construction of a nest? *Anim. Behav.*, **37**, 210–14.

ILAR (1980) *ILAR News*, **Vol. 23, No. 2–3**, National Academy Press, Washington, DC.

Jaeckel, J., (1989). Benefits of social contact and training in a primate unit, in *Laboratory Animal Welfare Research, Primates*, pp. 23–5, Proceedings of a UFAW Symposium, UFAW, Potters Bar.

Jeppeson, L.E. (1981) An artificial sow to investigate the behaviour of suckling piglets. *Appl. Anim. Ethol.*, **7**, 359–67.

Jones, R.B. (1987a) The assessment of fear in the domestic fowl, in *Cognitive Aspects of Social Behaviour in the Domestic Fowl* (eds R. Zayan and I.J.H. Duncan), pp. 40–81. Elsevier, Amsterdam.

Jones, R.B. (1987b) Social and environmental aspects of fear in the domestic fowl, in *Cognitive Aspects of Social Behaviour in the Domestic Fowl* (eds R. Zayan and I.J.H. Duncan), pp. 82–149, Elsevier, Amsterdam.

Lewis, N.J., Hurnik, J.F. and Gordon, D.J. (1982) Nursing apparatus for neonatal piglets. *Can. J. Anim. Sci.*, **62**, 975–8.

McEwen, P. (1986) *An Artificial Termite Mound for Orang Utans*, UFAW, Potters Bar.

McLean, K.A., Baxter, M.R. and Michie, W. (1986) A comparison of the welfare of laying hens in battery cages and in a perchery. *Res. Dev. Agric.*, **3**, 93–8.

Markowitz, H. (1982) *Behavioural Enrichment in the Zoo*, Van Nostrand Reinhold, New York.

Martin, P. (1989) Psychoimmunology: relations between brain, behaviour and immune function, in *Whither Ethology?* (eds p. Bateson and P. Klopfer), *Perspectives in Ethology*, **8**, 173–214.

Miller, N.E. and Kesson, M.L. (1952) Reward effects of food via stomach fistula compared with those of food via mouth. *J. Comp. Physiol. Psychol.*, **45**, 555–64.

Mills, A.D. and Wood-Gush, D.G.M. (1982) Genetic analysis of frustration responses in the fowl. *Appl. Anim. Ethol.*, **9**, 88–9.

Murphy, C. (1988) The 'new genetics' and the welfare of animals. *New Scient.*, 10 Dec. 1988, 20–1.

Murphy, L.B. (1977) Responses of domestic fowl to novel food and objects. *Appl. Anim. Ethol.*, **3**, 335–49.

Murphy, L.B. (1978) The practical problems of recognising and measuring fear and exploration behaviour in the domestic fowl. *Anim. Behav.*, **26**, 422–31.

Murphy, L.B. and Wood-Gush, D.G.M. (1978) The interpretation of the behaviour of domestic fowl in strange environments. *Biol. Behav.*, **3**, 39–61.

Nash, V.J. (1982) Tool use by captive chimpanzees at an artificial termite mound. *Zoo Biol.*, **1**, 211–21.

Oldfield-Box, H. (1981) Domestication, in *The Oxford Companion to Animal Behaviour* (ed. D. McFarland), pp. 136–9, Oxford University Press, Oxford.

Paxman, P. (1981) Quantock loose-housed system for veal calves, in *Alternatives to Intensive Husbandry Systems*, pp. 95–102; UFAW, Potters Bar.

Poole, T.B. (ed.) (1987) *The UFAW Handbook on the Care and Management of Laboratory Animals* (6th edn), Longman Scientific, Harlow.

Poole, T.B. (1988a) Behaviour, housing and welfare of non-human primates, in *New Developments in Biosciences: Their Implications for Laboratory Animal Science* (eds A.C. Beynen and H.A. Solleveld), pp. 231–7, Martinus Nijhoff, Dordrecht.

Poole, T.B. (1988b) Normal and abnormal behaviour in captive primates (Paper given at the XIIth Congress of the International Primatological Society), *Primate Report*, **22**, 3–12.

Reinhart, V., Hauser, D., Eisele, S. *et al.*, (1988) Behavioural responses to unrelated rhesus monkey females paired for the purpose of environmental enrichment. *Am. J. Primatol.*, **14**, 135–40.

Rooijen, J. van (1983) Operant preference tests with pigs. *Appl. Anim. Ethol.*, **9**, 87–8.

Royal Society/UFAW (1987) *Guidelines on the Care of Laboratory Animals and their Use for Scientific Purposes.* Part I. Housing and Care, RS/UFAW, Potters Bar.

Sales, G. and Wilson, K.J. (1986) *Sources of Ultrasound in the Laboratory*, U.F.A.W., Potters Bar.

Sales, G., Evans, J., Milligan, S. and Langridge, A. (1989) Effects of environmental ultrasound on laboratory rats, in *Laboratory Animal Welfare Research, Rodents*, pp. 7–16, Proceedings of a UFAW Symposium, UFAW, Potters Bar.

Savory, C.J. and Duncan, I.J.H. (1982) Voluntary regulation of lighting by domestic fowls in Skinner boxes. *Appl. Anim. Ethol.*, **9**, 73–81.

Stolba, A. (1981) A family system in enriched pens as a novel method of pig housing, in *Alternatives to Intensive Husbandry Systems*. pp. 52–67, UFAW, Potters Bar.

Stolba, A. and Wood-Gush, D.G.M. (1981) The assessment of behavioural needs of pigs under free range and confined conditions, *Appl. Anim. Ethol.*, **7**, 388–9.

Tauson, R. (1980) Cages: how could they be improved? in *The Laying Hen and its Environment* (ed. R. Moss), pp. 269–99, Martinus Nijhoff, The Hague.

UFAW (1988) Why zoos? *UFAW Courier*, No. 24, UFAW, Potters Bar.

UFAW (1989) *Laboratory Animal Welfare Research*, Proceedings of a UFAW Symposium, UFAW, Potters Bar.

Vestergaard, K. (1980) The regulation of dustbathing and other behaviour patterns in the laying hen: a Lorenzian approach, in *The Laying Hen and its Environment* (ed. R. Moss), pp. 101–20. Martinus Nijhoff, The Hague.

Walker, S. (1983) *Animal Thought*, Routledge and Kegan Paul, London.

Wallace, M.E. (1984) The mouse: in residence and in transit, in *Standards in Laboratory Animal Management*, pp. 25–39, Proceedings of a LASA/UFAW Symposium, UFAW, Potters Bar.

Webster, J. and Saville, C. (1981) Rearing of veal calves, in *Alternatives to Intensive Husbandry Systems*, pp. 86–94, UFAW, Potters Bar.

Wegner, R.-M., Rauch, H.-W. and Torges, H.-G. (1981) Choice of production systems for egg layers. *Proc. I Europ. Symp. Poult. Welfare*, Koge, pp. 141–8, W.P.S.A., Copenhagen.

Wiepkema, P.R. (1985) Abnormal behaviours in farm animals: ethological implications. *Neth. J. Zool.*, **35**, 279–99.

Wood-Gush, D.G.M., Dawkins, M. and Ewbank, R. (eds) (1981) *Self-awareness in Domesticated Animals*, UFAW, Potters Bar.

8

Behavioural problems in companion animals

VALERIE O'FARRELL

By convention, in veterinary medicine, pets are divided into two cate-gories: dogs and cats on the one hand and 'exotics' on the other. It seems odd to include with alligators and tarantulas pets such as rabbits or goldfish, but the distinction does embody an important difference. Dogs and cats are normally the only pets which are expected to share, at least to some extent, the owner's living space. The behavioural considerations to be taken into account in their maintenance are therefore not the same as those which apply to animals confined to their own cage or tank. Accordingly, in this chapter, the two categories will be dealt with separately.

8.1 DOGS AND CATS

Dogs and cats have not been singled out by chance for their special companionship role. They are particularly suited for it because of their general behavioural adaptability, their ability to become house-trained and their capacity to form social attachments to human beings. Although of course abuse and ill-treatment of dogs and cats does occur, it excites the condemnation of society in a way that the ill-treatment of exotic pets does not. It is generally accepted that a dog or a cat should not be cooped up alone in a boring environment and the deliberate encouragement of intra-species aggression is in some instances (e.g. dog-fighting) illegal.

Most 'behavioural problems' which dogs and cats develop are problems because they make the animal difficult to live with, not because they cause the animal distress. However, they are often indirectly detrimental to the animal, because they may lead to the owner abandoning it or having it destroyed. Two of the most common behaviour problems in dogs are aggression and destructiveness. Of the cases with these diagnoses seen at the Small Animal Clinic of the Royal (Dick) School of Veterinary Studies before specialist treatment was available, one-third were eventually destroyed. Where behaviour problems are tolerated, this may lead to a reduced quality of life for the animal and the owner. For example, dogs which tend to fight with other dogs may never be allowed off the lead outside.

Such problems are not rare, in dogs at any rate. Wilbur (1976) in a telephone survey of 350 American Owners found 25% to be dissatisfied with the behaviour of their dogs. O'Farrell (1987) found a similar proportion (10 out of 50) of Edinburgh dogs showing serious behavioural problems such as aggression. A further 3 in 5 showed less serious problems which caused the owners some inconvenience, such as over-excitement with visitors or occasional urination or defaecation in the house.

Similar figures are not available for cats, but judging by the smaller number referred to animal behaviour specialists, fewer cats are behaviourally unsatisfactory. This may be because they tend to lead lives which are more independent of their owners and less is expected of them. A cat owner whose cat steals unguarded food from the table tends to be philosophical, whereas many dog owners are outraged if their pet does the same thing.

It is only in the last few years that treatment has been available for behavioural problems. In 1974 Tuber *et al.*, pointed out that experimental findings concerned with animal learning could be applied to the clinical field. Since that time the number of animal behaviourists specializing in problems of dogs and cats has gradually increased, using concepts and findings from various branches of psychology and zoology, e.g. ethology, developmental psychology, psychology of stress and anxiety and human clinical psychology.

Each situation in which a dog or cat's behaviour has become a problem to its owner is a unique product of a particular combination of factors. The various factors which may contribute to the development of behaviour problems are discussed below, together with their relevance for prevention. This is followed by an account of treatment methods available. Finally, a few of the more commonly occurring behaviour disorders are discussed, together with suggestions as to how to manage their behaviour.

8.1.1 Causes of behavioural problems

Genetic

Genetic factors may make a contribution to a range of behavioural problems. There is evidence that general behavioural traits, such as nervousness or hyper-reactivity, may be inherited. Over a single generation Murphree *et al.*, (1967) developed by selective breeding two strains of pointers, one 'stable' and the other 'neurotic', whose behaviour differed significantly on experimental tests. The raised incidence of particular behaviour problems in certain breeds also suggests that more specific responses may be inherited. For example, Siamese cats are particularly prone to the habit of sucking or chewing knitted garments; many Border collies show a fear reaction to the sight of people waving things.

The implications of these findings for prevention seem clear: breeders should not breed from affected animals. However, the temptation to do so is sometimes overwhelming when the animal's physical attributes win prizes at shows. Many behavioural tendencies are not evident in the show ring. Moreover the position with regard to some behavioural traits is more complex. In some breeds, traits which are potentially problematic are selectively bred for: for example, aggression in guard dog breeds.

Potential owners can to some extent protect themselves from problems by selecting a breed with a relatively low incidence of the kind of behaviour they particularly wish to avoid. It can be surprisingly difficult, however, to discover what problems are typical of a breed. Breeders tend to view their chosen breed through rose-coloured spectacles. Books about different breeds often need to be interpreted in the same way as estate agents' literature. For example, in a dog for 'independent minded' read 'inclined to show dominance aggression'. The foregoing, however, should not be taken as an argument for acquiring mixed-breed rather than pure-bred animals. Although the eventual temperament of a pedigree puppy or kitten is not as predictable as its physical appearance, it is more predictable than that of a mongrel.

Hormonal factors

The incidence of certain behavioural problems is raised in male dogs and tom-cats. In male dogs, dominance aggression directed towards human beings and other dogs, roaming and urine-marking are more common than in bitches. In male cats, roaming, urine-marking and intra-species aggression are more common than in females and these types of behaviour are even more common in tom-cats than in castrated males. There is also evidence that spaying of bitches in some cases exacerbates dominance aggression (O'Farrell, 1989).

Other things being equal, therefore, one might recommend to owners who wish to minimize the likelihood of behavioural problems that they acquire female dogs or cats. Owners, however, often have a personal preference for a male animal. They may also not want the expense or inconvenience of preventing unwanted pregnancies in the female.

Castration of tom-cats, as well as greatly reducing the probability of the problems outlined above, usually results in improvement in animals already showing the problem behaviour.

Castration of male dogs tends to reduce or prevent roaming, but has a much less reliable effect on dominance aggression, sexual mounting and urine marking, producing improvement in only 50–60% of cases (Hopkins *et al.*, 1976). This may be because the brain of the male puppy is affected by hormones circulating perinatally (Hart and Ladewig, 1979). If castration of a dog is being considered, it is wise to test its probable effect by first administering a drug such as delmadinone which produces the same effect pharmacologically. On the present evidence, bitches which show evidence of dominance aggression should, if possible, not be spayed.

Synthetic progestagens may be useful in the treatment of a wide range of disorders, megestrol and medroxyprogesterone being the drugs most often used. They may reduce the frequency of urine spraying in castrated male cats and females. They may also be a useful adjunct in the behavioural treatment of dominance aggression in dogs, usually being more effective than castration.

Developmental factors

There is clear evidence of a sensitive period in the lives of dogs and cats which lasts from 2–3 weeks until about 12 weeks of age. Kittens and puppies which have no exposure to human beings during that time will never become properly tame. There is also evidence of more subtle defects in animals which have restricted interaction with people. Scott and Fuller (1965) found that puppies reared in a home environment were generally more confident with human beings than those reared in kennels.

The same principle applies to exposure to a range of situations. Puppies separated too soon from their mothers and litter-mates (i.e. before 6 weeks) may as adults show deficiencies in interactions with other dogs. Males may be unable to copulate effectively, attempting to mount from the side or front. Either sex may get into fights with other dogs, not because they are abnormally aggressive, but because they do not display the appropriate submissive behaviour in response to a dominance threat.

Kittens and puppies may also become acclimatized to other species during this sensitive period. If introduced to each other (i.e. dogs to cats), they will tend to co-exist amicably thereafter. This lack of fear can, of course, have disadvantages, as when a cat makes no attempt to escape from a strange predatory dog.

The underlying mechanism of this sensitive period in puppies was investigated by Scott and Fuller (1965). They found that a puppy's tendency to approach new people and things was strongest at about 5 weeks of age, but declined thereafter as its fear of novel stimuli increased. At around 6–8 weeks, therefore, a puppy can be relatively easily habituated to a range of stimuli, to prevent inconvenient fears developing later. These stimuli should include a range of domestic sights and sounds (television, vacuum cleaner, telephone, etc.) and people of varying ages and sexes. Many attacks by dogs on children and babies (occasionally with fatal results) might be avoided if the dogs were exposed to this age group during the sensitive period. Many dogs become unsettled when travelling in cars. This reaction may be prevented by taking young puppies for car rides. Similarly, some dogs show a fear of traffic or busy streets. Owners are understandably cautious about exposing puppies to these situations before they are fully immunized at 16 weeks, but small breeds can be carried to the shops or to the post in the owner's arms.

Perhaps the most important precaution an owner can take against developmentally induced behaviour disorders is to buy a puppy or kitten which has been reared in a psychologically optimal environment. It should be acquired from 8–12 weeks of age from a reputable breeder. From 3 weeks of age it should have spent a large proportion of the day in the household, not isolated in a kennel or cattery. Puppies or kittens should not be bought from dealers or pet shops, both because the circumstances of their early life are unknown and because they have inevitably suffered stressful changes of environment.

Even after the so-called 'socialization period', traumatic events can have lasting adverse effects. People who acquire dogs rejected by previous owners frequently attribute their fearfulness or aggression to maltreatment by these owners, but these suppositions are usually hard to confirm. On the other hand, there is clear evidence that dogs acquired from animal shelters are more likely to become anxiously over-attached to their new owners (McCrave and Voith, 1986). This can result in the extremely inconvenient problem of stress-induced destructive activities (chewing loose objects, scratching paintwork, etc.) when left alone in the house. Owners who acquire dogs from this source should therefore be warned of this possibility and instructed in treatment methods (see below).

Inconvenient behaviour patterns

Often an animal's species typical behavioural repertoire can cause problems, merely because it is performed in an inconvenient situation. An example of this is predatory aggression in both dogs and cats. Although some cat-owners may value their pets' mousing abilities, others may be upset by finding dead, or worse, dying small birds and rodents in the house. Predatory aggression in dogs may have even more undesirable results. Dogs may attack sheep, cats, bicycles, joggers and occasionally, with tragic results, babies. Obviously a dog which displays predatory aggression towards children or babies cannot be tolerated, but other forms of predatory aggression may be controlled in dogs by substituting an alternative set of learned responses (e.g. coming when called). The inconvenience of other instinctive behaviour patterns can often be similarly modified. Many dogs which are about to chase after a bitch in oestrus will come back to the owner when called; others will not.

In cats, the only feasible method of treatment is usually to divert the behaviour into more acceptable channels, by providing another stimulus which triggers it. For example, cats can be prevented from sharpening their claws on the furniture by providing a scratching post which embodies as many attractive features as possible, e.g. a vertical surface, covered in a loose weaved material. In cats, predatory aggression directed against birds or rodents cannot normally be modified, but when the aggression is directed playfully towards human beings (pouncing on feet, hands, etc.) it may be diverted onto toys.

Misunderstanding social behaviour

Dogs and cats are social animals, with the ability to form attachments to human beings. They communicate this attachment by means of signals which are similar enough to those of human beings to enable communication to take place between the two species. This is what makes ownership of a cat or dog gratifying in ways in which ownership of, say, a goldfish, can never be.

On the other hand, there are differences in what human beings normally expect in a social relationship and what dogs and cats instinctively expect. Some social signals are also so different from those used by human beings that their meaning cannot be intuitively grasped. In the case of the cat, these differences are normally obvious and are usually accepted by cat owners as part of its 'inscrutable nature'. If a cat which has been happily sitting on its owner's knee suddenly jumps down and runs off for no apparent reason, the owner may feel slightly rejected, but normally this kind of incident does not lead to problems. On the other hand, dogs and their owners frequently become enmeshed in a pernicious web of misunderstanding.

The most frequent source of this kind of misunderstanding is the relationship of dominance. As dogs reach adulthood (normally at about 2 years old) many begin to construe their relationship with fellow pack members (i.e. the family in which they live) in terms of dominance; some begin to do this much earlier in life. Some (mostly male dogs and those selectively bred for dominance aggression) seem more preoccupied with dominance than others. Such a dog may form an opinion as to his position in the dominance hierarchy as a result of testing out the responses of his fellow pack members. The significance of some of these tests (e.g. growling when another family member approaches his food bowl or bed) may be obvious to the owners but the significance of other kinds of behaviour may escape them. For example, if a dog makes frequent requests of the owner (e.g. for attention, for a ball to be thrown, to be let out of doors) and the owner complies, the dog may conclude that the owner has assumed a subordinate role. The problems arise when the owner, having become, in the dog's eyes, established in the subordinate position, behaves towards the dog in a way which it construes as inappropriately dominant. Examples of such behaviour by the owner might be patting the dog, attempting to groom it, reprimanding it or attempting to move it from its resting place. The dog may respond to such a perceived dominance challenge with aggression which is to the owner often unexpected and inexplicable.

Such problems can be prevented if the owner is aware of when a dog is becoming preoccupied with dominance and does not allow itself to be assigned to a subordinate role. They can also be treated if the owner is instructed as to how to interact with the dog in order to establish itself in the dominant role. He or she should ignore all the dog's social initiatives: requests for food, attention, play, etc. The owner should use these incentives to reward the dog only when it obeys a command.

Faulty learning

Problem behaviour can be learned, either by classical conditioning or operant learning. For example, dogs and cats may learn by classical conditioning to urinate or defaecate in the house. A cat may initially urinate on the carpet because it has been accidentally shut in the room. If this happens repeatedly, the act of urinating may become associated with the texture of the carpet and the smell of the old urine there.

Dogs are particularly liable to acquire problem behaviour via operant learning. Owners commonly do not appreciate the power of social reinforcement. The behaviour of a dog which is generally hyperactive or which engages in various persistent and annoying activities such as begging for food, barking or running off with its owners possessions, may be maintained by the owner paying it attention when it is thus engaged, but ignoring it when it is quiet and well behaved.

Problem behaviour may also be learned because of the inappropriate use of punishment. Dogs may, for example, associate the punishment with a feature of the situation other than that intended by the owner. This commonly happens during the house-training of puppies. An owner may punish a puppy on discovering it urinating indoors; the puppy may associate the punishment with the presence of the owner. On another occasion, therefore, when the owner escorts the puppy outside to urinate, it may not oblige, but immediately they return to the house it may go away from the owner into another room and urinate there.

Once the reinforcements, both positive and negative, which are maintaining a piece of behaviour have been identified, treatment consists of appropriate alteration of these reinforcement contingencies. Because of the unpredictable effects of punishment, it should not be used, if possible. Undesirable behaviour should simply be followed by no reinforcement, as when a dog pestering for attention is ignored. Alternatively, the undesirable behaviour may be prevented before it occurs by distraction or by substitution of another response: a dog which is liable to bark uncontrollably when visitors arrive may be called to the owner's side and made to sit as soon as it pricks up its ears at the sound of footseps coming up the path. Desirable behaviour should, of course, be rewarded; for example, a restless, overactive dog should be praised when it is calm.

Stress and anxiety

Stress increases the likelihood of various kinds of problem behaviour. In dogs, the most common of these are destructive chewing, sexual mounting of people or inanimate objects, self-mutilation and stereo-typic dashing or pacing. In cats, the commonest problem activities are urine marking and self-mutilation.

Many dogs are stressed by separation from their owners: dogs which have a previous history of traumatic separation are likely to be particularly upset. For cats, the addition of human or animal members to the household is liable to prove more stressful.

Punishment can increase anxiety and it is therefore particularly important that owners do not punish stress-induced activities in either species. In addition, dogs in particular may be put into a state of conflict by being unpredictably rewarded and punished for the same behaviour. This may be the result of inconsistency on the owner's part: he or she may, for example, welcome the dog's social overtures when feeling relaxed but may become infuriated by them when busy. It may also be the result of differing reactions among family members to the dog's behaviour. For example, some might welcome the dog onto the sofa beside them, whereas others might shout at it and push it off. Such inconsistencies may be the by-product of upset or crisis in the family, which may be why problem behaviour in cats and dogs often appears at times of general family disturbance. Treatment in these cases obviously involves eliminating the source of stress where possible.

Both cats and dogs may display fear reactions (avoidance, cowering, trembling, etc.) to specific stimuli. Not uncommonly, this is accompanied by aggression, which may be directed against human beings. In some cases, it may be clear that the fear was learnt by association with a traumatic event. Sometimes a deprived early en-vironment seems to be a contributory factor. Often, however, the origin of the fear is obscure.

Whatever the cause, systematic desensitization is the treatment of choice. This treatment is widely used in the treatment of human phobias, but is at least as effective in the treatment of animals. In essence, it involves arranging that the patient perform a response incompatible with anxiety in the presence of the fear-provoking stimulus. In the case of dogs and cats, the response incompatible with anxiety is usually sitting calmly being petted by the owner and being fed tit-bits. While in this calm state, the animal is presented with a mild version of the fear-provoking stimulus. For example, if a dog is afraid of thunder-storms, a tape-recording of thunder might be played at low volume. If the animal is able to tolerate the mild version of the stimulus without anxiety, the intensity of the stimulus is gradually increased. If the animal shows any evidence of anxiety, this is a sign that the intensity of the stimulus is being increased too quickly. During the period that the treatment is being carried out, the animal should not be exposed to a full-blown version of the phobic stimulus. For example, the treatment of a thunder phobia should be carried out in winter.

If carried out correctly, this treatment method is usually successful. The difficulty most often encountered is in devising a hierarchy of stimuli which increase appropriately in intensity and which can be presented under controlled conditions.

The use of systematic desensitization need not be confined to the treatment of learnt fear. It can also be used in disorders where a learnt association has been formed between specific stimuli and other states of high arousal, such as aggression or excitement.

Owner attitudes

Owners sometimes contribute to behaviour disorders because they try to form a relationship with their pets which is inappropriate. This happens more often with dogs than with cats, whose behaviour is less readily influenced by that of their owners.

Owners may, for example, treat a dog as a substitute for a baby, taking pleasure in gratifying its wants and complying with its requests. In many instances, this kind of relationship is pleasurable to both parties but it may in some cases lead to problems. For example, if the owner chooses to gratify the dog by feeding it on demand, there is a risk that the dog may over-eat and become obese. If the dog is inclined to dominance, as mentioned above, complying with its requests may lead to the development of dominance aggression. There is evidence that in dogs such aggression is indeed correlated with attitudinal factors in the owners, such as wanting love and affection from their dogs and feeding them food designed for human consumption (O'Farrell, 1987).

Owner anxiety has also been shown to be correlated with dog behaviour disorders, particularly those involving displacement activities (O'Farrell, 1987). The most likely explanation of this correlation seems to be that anxious owners are more prone to behave inconsistently towards their dogs, thus raising the dogs' own levels of stress. Alteration of owner attitudes is often difficult as these are frequently linked to long-standing personality attributes. Some change in attitude-linked behaviour, however, is sometimes possible if the following considerations are borne in mind.

Comments about the effect of the owner's attitude on a dog's behaviour should be neither judgemental (e.g. 'you spoil that dog') nor vague (e.g. 'you should be firmer with him') but should make factual observations about specific behaviour (e.g. 'It seems from what you tell me that you often reward the dog's barking by paying him attention'). This kind of comment has two advantages. It invites the owner to stand back and observe his or her own behaviour and thereby encourages him to review his whole attitude to the dog. It also offers specific instructions which, when followed, often themselves give rise to a change in attitude. For example, ignoring a dog for much of the time may reduce an owner's emotional involvement.

8.2.2 Treatment of common behavioural problems

Although, as discussed above, each individual case must be considered separately, it is often useful to bear in mind the most usual causes of the commoner behavioural problems and the treatment methods likely to prove effective.

Dogs

Aggression This can be subdivided into dominance and predatory aggression.

Dominance aggression Dominance aggression is a social communication directed towards presumed members of the same species (i.e. people and dogs). It is normally preceded by threats, expressed first by bodily posture (see Figure 4.2, Chapter 4), then by growls and attack. If the victim expresses submission by means of his bodily posture (Figure 8.1), the attack is normally aborted.

This kind of aggression may be directed towards family members who have allowed the dog to believe itself to be dominant over them (see above). The treatment in this case should consist of instructing the owners how they can establish themselves in the dominant role. The concurrent administration of megestrol for a few weeks is often helpful, as it usually induces temperamental changes which make the dog easier to dominate. To begin with, the owners should avoid situations which might provoke aggressive confrontation with the dog. Once they sense that their relationship with the dog has changed, they should begin systematically to desensitize it to the particular situations which previously provoked aggression (see above). For example, if the dog habitually guards his food bowl, he should first of all not be given a food bowl at all: he should be given all his food as 'tit-bits' conditional on obeying commands. When he has assumed a more subordinate role in the household, the food bowl should be gradually reintroduced; the dog might first of all be fed in a

(a)

(b)

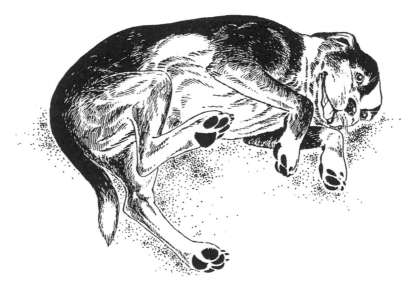

Figure 8.1 (a) Active submission; (b) passive submission.

different room from that to which he is accustomed and from a different dish.

It should be noted that this approach is radically different from that advocated by many dog handlers, which is to engage in a séries of confrontations with the dog, which are, if necessary, violent and result in decisive victory for the owner. There are several reasons why this line of action is inadvisable. Firstly, an owner who is established in a subordinate role may be put in danger by confronting a dominant dog. Secondly, even if the owner wins the confrontation, the episode may have all the after-effects of any kind of punishment. More particularly, it is likely to raise the emotional temperature of the owner/dog relationship and may well make the dog more determined to regain the dominant position on another occasion.

Aggression towards people outside the family (e.g. visitors, postmen, neighbours) is frequently accompanied by more subtle signs of dominance over family members: although he may never actually threaten them, the dog may tend to lead a life which is fairly independent from them, initiating most of the social interaction with them. This dominance should be treated as outlined above and again the dog systematically desensitized to the provoking stimuli. For example, a dog which attacks visitors might be made to sit calmly and under the owner's control while a stranger passed by the garden gate but did not enter. The next stage might be for a visitor to open the gate, but not come up the path and so on.

Aggression triggered by pain or fear (e.g. by a distressing veterinary procedure) is most usefully viewed as a variant of dominance aggression. To inflict pain on a dog may be viewed by it as a dominant act; dogs inclined to take up this challenge may react with aggression. Others may react with fear and a submissive posture. The treatment of this variant of dominance aggression is similar to that outlined above. The underlying dominance relationship should be altered and the dog systematically desensitized to the provoking stimuli. However, this procedure is inappropriate for comparative strangers such as veterinary surgeons who have to find some way of handling the dog immediately. Fortunately, many dogs who are well established in a dominant role *vis-à-vis* their owners will adopt a subordinate role *vis-à-vis* a stranger encountered in an unfamiliar place. If there is any doubt as to the dog's possible reaction, however, it should be appropriately restrained (e.g. with a tape muzzle) before any potentially threatening procedures are attempted.

The relative dominance status of dog and owner can also be a factor contributing to a range of problems where no aggression is manifest at all. Where the dog's refusal to respond to the owner's command leads to difficulties (e.g. if it will not return to the owner at the end of a walk), increasing owner dominance may be helpful.

Aggression towards other dogs may involve dogs in the same household or dogs from different households. Where two or more dogs in the same household are involved, the cause is usually a struggle for dominance. This may be precipitated by hormonal changes in one dog, for example by its reaching sexual maturity or giving birth to a litter. It may also be encouraged by owners punishing the dominant dog and favouring the subordinate one, thus giving the subordinate one ideas above its station. Sometimes the situation is further complicated by different family members favouring different dogs. The remedy is to help the dogs towards a firm agreement as to their relative positions by reinforcing their dominant and subordinate roles. This can be done behaviourally by giving preferential treatment to the dominant dog. Concurrent hormonal manipulation may be helpful.

Aggression towards strange dogs encountered on walks is harder to treat as the behaviour of the strange dog cannot be manipulated. It is often helpful to increase the dominance of the owner over the dog and therefore the degree of control which can be exercised. Hormonal manipulation may also be useful and, if it is possible to arrange, systematic desensitization to strange dogs.

Predatory aggression   Predatory aggression is directed towards presumed animals of a different species. Common victims are sheep, cats, smaller dogs, bicycles, joggers and babies. These attacks are often triggered by vocalization and quick movement of the victim. They are not preceded by threats and a victim cannot prevent them by taking up a submissive posture. If he cannot fight off the dog or escape, a victim's best strategy is to lie motionless and quiet.

This behaviour problem is hard to treat, although increased control by the owner and systematic desensitization may reduce it to manageable proportions. This is one situation where punishment is worth trying, especially if the alternative is euthanasia. For example, successes have been reported using shock collars with sheep-chasers.

Destructiveness in owner's absence   Because of the amount of damage a dog can do when left alone in a house or car, this is a problem which most owners cannot tolerate.

The commonest cause of this behaviour is the dog's agitation at being left alone. The destructive activities are probably most appropriately viewed as stress-induced activities although some of these (e.g. scratching at the door with fore-paws) are probably directed towards removing the barriers which separate the dog from its owner.

The dog may at the same time engage in other activities which present a problem. It may whine or bark continuously, attracting complaints from the neighbours. Because of increased autonomic activity it may urinate and defaecate indoors.

The most severe cases usually occur when the owners are out of the house for a few hours, but the problem can also occur when the dog is shut away from the owners. Agitation at separation is a common cause of the dog fouling the house at night.

There are various factors which may predispose a dog to developing this kind of problem. One is a history of traumatic separation in early life. Another is suddenly being left for long periods, having previously being accustomed to the owner's constant presence; as a preventative measure it is wise, therefore, to accustom dogs from puppyhood to separations of increasing length. Separation anxiety is also often linked with an intense relationship with the owner, who may be emotionally dependent on the dog. This relationship is often characterized by some ambivalence on both sides; on the part of the dog this ambivalence is often increased by the punishment meted out by the owners on discovering the damage!

In a few cases, the problem is easily solved once its cause is explained to the owner. For example, if a dog fouls or is destructive when shut in the kitchen at night, this behaviour usually ceases if the dog is allowed access to the owner's bedroom. However, the majority of these problems are not resolved so easily, usually because the owner cannot or will not stop leaving the dog alone.

In these cases, the first aim should be to reduce the emotional intensity of the relationship between dog and owner. The owner should be persuaded to stop punishing the dog. Some owners are convinced that the dog 'knew he had done wrong', because of his guilty look when the owner returns. It should be explained to them that this 'guilt' is merely fear of anticipated punishment and that this punishment, administered a long time after the crime, can have no deterrent effect. The owner should also be encouraged to behave more distantly towards the dog, ignoring it much of the time, especially before and after leaving it alone. It should be trained to spend time in a different room from the owner, by rewarding it for tolerating separations of increasing length, without actually shutting the door between them.

When the dog appears to be less dependent on the owner, it should then be systematically desensitized to actual departures, starting with a pre-departure routine and then moving on to very short absences (i.e. 10 seconds). The length of the absences should then be gradually increased.

Ideally, over the period that this treatment is being carried out, the dog should not be subjected to the kind of separation which normally provokes destructiveness. This may mean carrying out the treatment during a holiday from work or making other arrangements for the dog. Where this is not possible, it may be necessary to prescribe a tranquillizing drug. For dogs and spayed bitches, megestrol is sometimes a useful alternative: as well as reducing dominance, it tends to reduce the activity level of some dogs.

Cats

Inappropriate urination and defaecation   This is by far the most common behaviour disorder occurring in cats. In these cases it is necessary to differentiate urine marking from normal urination. In urine marking, small amounts of urine are deposited. The cat usually stands in a characteristic posture (see Figure 8.2) and sprays onto a vertical surface. In normal urination, the cat adopts a squatting posture and larger quantities of urine are involved. Occasionally a cat may mark in a squatting posture, the distinguishing feature being the small amount of urine deposited.

Urine marking   This is probably some form of territorial behaviour, although it sometimes seems to be performed as a stress-induced activity. It is most commonly shown by tom-cats and castration is the

Figure 8.2 The characteristic posture of a cat spraying.

treatment of choice. However it is also seen in females and castrated males, in which cases synthetic progestagens may prove helpful.

Attention should also be paid to possible psychological factors. Spraying is commoner in households with more than one cat. If other measures fail, therefore, finding one of the cats another home may resolve the problem in both of them. Spraying may also be prompted by the arrival of another cat in the neighbourhood or by the addition of another human or animal member to the household. If the additions are only temporary it is worth waiting to see if the spraying dies down when the visitors leave. Otherwise some territorial rearrangement, such as providing the cat with a place where it cannot be disturbed, may ease matters.

Spraying may be seen when the animal is frustrated or under stress. For example, cats may spray if their meal is not served on time or if they are shooed off the kitchen table. If this kind of incident seems to be the trigger, with some forethought it may be possible to prevent the situations from occurring. Punishment is contraindicated as it tends to increase stress.

If a cat normally uses a litter tray but begins to urinate or defaecate deter it by putting its food dish there, as cats prefer to keep the two functions separate. Alternatively, it may be discouraged by aluminium foil covering the area normally sprayed. The high-pitched noise produced by spraying on it seems to be aversive. The smell of urine should as far as possible be eradicated, using bleach or a biological detergent. Disinfectants with a strong smell should be avoided as they tend to attract the cat back to the same spot.

Normal urination and defaecation   When a cat urinates or defaecates normally but in a place not approved by the owner, a different approach is required. The owner must first try to discover what is deterring the cat from using its customary toilet areas. For example, the cat may need to urinate or defaecate with increased frequency or urgency. It is worth investigating the possibility of physical causes, especially if there are other physical signs such as increased water intake. Cats which become less active, either from illness, pregnancy or old age, may stop making the effort to go outside or to visit a litter tray in a distant part of the house.

If a cat is active indoors, but reluctant to go outside for any purpose, it may be afraid to go out and it is worth investigating possible sources of fear, e.g. neighbouring children or cats.

If a cat normally uses a litter tray but begins to urinate or defaecate in another part of the house, the litter tray may have become aversive in some way. The litter may not be changed often enough or the owner may have started to use a new kind of litter which the cat finds unacceptable. Alternatively the cat may feel vulnerable when using its tray, perhaps because he was startled when using it. Trays are now commercially available with covers which provide more privacy.

Even when the cause of the inappropriate urination or defaecation has been discovered and dealt with, the habit may still persist. This is usually because the cat has developed a preference for the new substrate, usually a soft one, such as a carpet or bed. The first approach to this problem is to deny the cat access to the place in the house he has become accustomed to using. When the habit of urinating and defaecating in the proper place has become re-established, the cat can be gradually allowed access to the forbidden areas, at first under supervision. As with urine spraying, traces of urine smell should be eradicated as far as possible.

In cases where the cat shows a reluctance to resume his former habits, it may be necessary to make the correct toilet area more attractive by incorporating into it significant features of the forbidden location. For example, if a rug has become saturated with urine, the best solution might be to put a portion in the flower-bed or litter tray, gradually covering it with soil once the cat's habit of urinating there is re-established.

8.2 EXOTIC PETS

It is not possible in the scope of this chapter to outline the particular ethological requirements of all the various species kept as pets. The best sources of information for the commoner 'exotics' such as the rabbit, gerbil, etc., are reputable pet books such as the series published by the RSPCA, but it should be borne in mind that there is still room for further observation of the species in question and devising methods for improving living conditions. In the case of the less common 'exotics', information about the animal's behaviour patterns when living in natural conditions can be harder to find.

As a general guideline, in determining the behavioural needs of a species, the following considerations should be borne in mind.

Physical environment

Commercially available housing is frequently deficient in complexity, providing too few hiding places or opportunities for exploration. Too frequently, a species cannot engage in its own species typical behaviour patterns. For owners, space is often at a premium, but a better environment does not necessarily take up more space. For example, the Mongolian gerbil is behaviourally adapted to digging tunnels and living in burrows. If housed in a bare cage it will engage in stereo-typies such as scrabbling repetitively in corners or bar-chewing. If, on the other hand, an aquarium tank of similar volume is filled with a mixture of peat and straw or cardboard tubing the animal will engaged in non-stereotypic tunnelling.

Social environment

Owners need precise information about the social behaviour of the species they plan to keep, in order to acquire the appropriate number and gender of animals. Social species kept singly may become ab-normally inactive or engage in stress-induced activities whereas housing territorial or solitary species together in too close confine-ment may result in dangerous intra-species aggression. For many of these latter species, the conventional solution is to house individuals singly, only allowing them to meet for breeding purposes. However, in some cases it is possible to devise an environment where individuals are able to manage their degree of mutual contact themselves. Rift valley cichlids, for example, are best kept in a group in a large tank which should have a complex rocky environment providing many hiding places.

Tameness (i.e. absence of fear or aggression towards owner)

Most mammals can be kept tame by frequent handling from an early age. This is advisable for several reasons. It enhances the pleasure of keeping the animal. It reduces the stress produced by unavoidable disturbance and human contact (e.g. when cleaning out the cage). It makes veterinary treatment easier. It reduces the probability of the owner being attacked by the animal in self-defence.

Owner education

Pet keeping differs from the husbandry of domestic animals in that, as far as catering for the behavioural needs of pets are concerned, there should be little conflict between the animal's interest and those of the owner. Pets kept in unsuitable conditions tend to be abnormally inactive or engage in stereotypies. They may also harm each other; adults may fight or mothers kill their young. These phenomena tend to upset owners, while an animal engaging in a wide behavioural repertoire can provide interest and pleasure. The main reason why so many pets live in ethologically unsuitable environments seems to be the ignorance of their owners as to their needs and how to meet them. Owners are not helped by the design of much commercially available housing or by the paucity of informed advice available from many pet shops. To remedy matters a sustained campaign of public education seems to be needed.

Pets are often owned by young children. However, the concept of a creature which has psychological needs which are radically different from, but as valid as one's own, is a relatively sophisticated one. Many young children treat pets like inanimate toys or treat them in inappropriately anthropomorphic ways. Close supervision is therefore needed until a child has reached the requisite level of maturity.

REFERENCES

Hart, B.L. and Ladewig, J. (1979) Serum testosterone of neonatal male and female dogs. *Biol. Reprod.*, **21**, 289.

Hopkins, S.G., Schubert, T.A. and Hart, B.L. (1976) Castration of adult male dogs: Effects on roaming, aggression, urine marking and mounting. *J. Am. Vet. Med. Assoc.*, **168**, 1108.

McCrave, E.A. and Voith, V.L. (1986) Correlates of separation anxiety in the dog. Paper given at Delta Society International Conference, Boston.

Murphree, O.D., Dykman, R.A. and Peters, J.E. (1967) Genetically determined abnormal behaviour in dogs; results of behavioural tests. *Conditioned Reflex.*, **2**, 199–205.

O'Farrell, V. (1987) Owner attitudes and dog behaviour problems. *J. Small Anim. Pract.*, **28**, 1037–45.

O'Farrell, V. (1989) *Problem Dog: Behaviour and Misbehaviour*, Methuen, London.

Scott, J.P. and Fuller, J.L. (1965) *Genetics and the Social Behaviour of the Dog*, University of Chicago Press, Chicago.

Tuber, D.S., Hothersall, D. and Voith, V.L., (1974) Clinical animal behaviour: a modest proposal. *Am. Psychol.*, **29**, 762–6.

Wilbur, R.H. (1976) Pet ownership and animal control, social and psychological attitudes. 1975 report to National Conference on Dog and Cat Control, Denver, Colorado.

Index